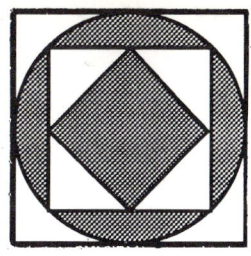

Maths workshop

Anita Straker

CAMBRIDGE
UNIVERSITY PRESS

Published by the Press Syndicate of the University of Cambridge
The Pitt Building, Trumpington Street, Cambridge CB2 1RP
40 West 20th Street, New York, NY 10011-42, USA
10 Stamford Road, Oakleigh, Victoria 3166, Australia

© Cambridge University Press 1992

First published 1992

Printed in the United Kingdom by Bell and Bain Ltd., Glasgow

ISBN 0 521 43819 5

Contents

Notes for teachers

Maths workshop contains resources selected from the *Extension Activities* which form part of the Cambridge Primary Mathematics scheme. They are intended for use in secondary schools and are capable of supplementing any mathematics materials. They include a variety of practical tasks, games, puzzles and investigations, ranging from Level 3 to Level 6 of the National Curriculum. Some sheets could be used as a starting point for a class activity; others are more appropriate for individuals or small groups. Some would make suitable assessment tasks to be used at intervals throughout a course of study; others could be used for homework, extending work based on other materials.

The book contains a series of photocopiable masters. Since pupils will record in their workbooks their solutions to many of the problems and investigations, none of the sheets asks them to write on it. This means that most can be handed back to be used again. Occasionally, you may feel it would be better if pupils worked directly on the sheet and, in such cases, you can recommend that they do so.

Some of the sheets suggest the use of number cards or spotty paper. If these are not already available, master copies are provided on pages 110 and 111.

The sheets are linked to the number, algebra, shape and space and handling data attainment targets of the National Curriculum, noted by symbols in the top left corner of each sheet. For example, N3 indicates Level 3 in number, A4 indicates Level 4 in algebra, S5 indicates Level 5 in shape and space and D6 Level 6 in handling data.

All the sheets are linked to the attainment target on using and applying mathematics since they provide activities in which pupils can plan strategies and think logically. Your role in these circumstances needs to be one of standing back so that pupils learn to make their own decisions about what to do and how to do it. This does not, of course, imply that they are not given help if they get 'stuck'. It is more a case of asking questions to prompt their thinking, rather than giving them specific information or directions about what to do.

For example, in the initial stages of clarifying a problem you might ask:

- What is the problem asking you to do?
- What information are you given? What would be the best way to organise it?
- Would it help to draw a diagram?
- Can you get any more information?

At the stage of attacking the problem, you might recommend drawing on previous experience, using intuition or even trial and error, by asking:

- What did you do last time?
- Could you try a particular case?
- What would happen if you started with a simpler problem - two shapes rather than three, four numbers instead of five, a square instead of a rectangle. . . ?
- Would it help to work backwards?
- What do you know that won't work?
- Why not make a guess and see what happens?

Once a solution is underway, but not yet complete, you might ask:

- Have you used all the information?
- Would it help to put things in order?
- Do any bits go together?
- Would a graph help? Or a diagram?
- Is there a pattern? How could you use it?

Some of the problems ask pupils to 'investigate different ways of . . .' or 'find as many ways as possible of . . .' Almost all pupils will be able to discover one or two possibilities; the challenge for them is to know when to stop searching. Good questions to ask in these circumstances are:

- Have you found all the possibilities?
- How have you organised them?
- How did you set about counting them?
- How do you **know** that you have found them all?

Once a first solution has been found, you can ask 'What if . . .?' questions to broaden the scope of an investigation.

- What if there were more of them?
- What if you changed the numbers, changed the shape, changed the rules . . .?
- What if you started with a rectangle rather than a square?
- What if you could only use . . .?

For most activities in this book, plenty of time needs to be allowed. Problem solving often takes longer than expected, especially if alternative solutions are explored. Pupils should gain encouragement from the fact that in real life mathematicians do not usually solve a problem in the neat time slot between break time and lunch! If any problem is unresolved by the end of the session, then leave it as an unanswered question. Put it away until tomorrow, next week or even next year, and come back to it in due course. For busy teachers who wish to avoid sleepless nights, some sample solutions are provided at the back of the folder.

I hope that the pupils in your school enjoy working on these activities as much as I have enjoyed writing them.

Anita Straker
Spring 1992

Number puzzles 1

1 You need some squared paper. Use each of the numbers 1 to 8.

Put one number in each blank square. Each line must add up to 12.

12 in a line

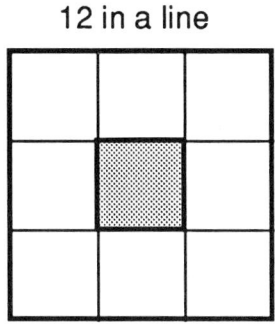

Now make 13 in each line. Then 14. Then 15.

13 in a line	14 in a line	15 in a line
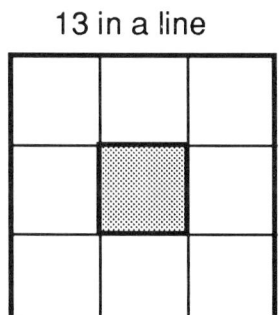	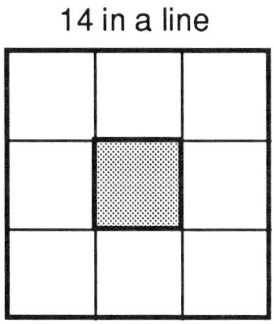	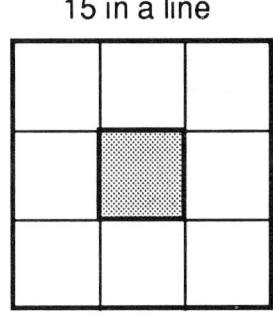

2 You need squared paper and 20 counters.

The first grid has 6 counters in each line, some in each box.

Arrange all the counters. Put 7 in each line. Put 8 in each line.

6 in a line	7 in a line	8 in a line
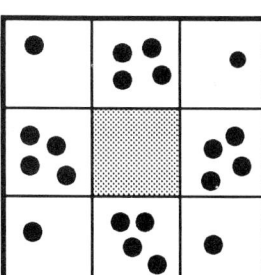	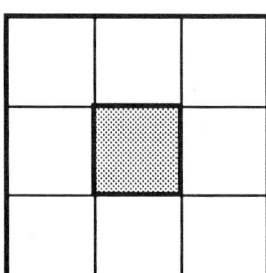	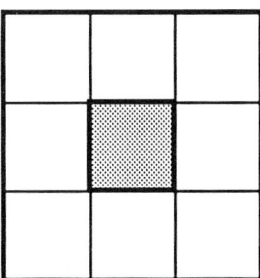

Record as many different ways as possible.

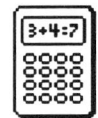

Number puzzles 2

1 You need your calculator. Use only these keys.

$$2 \quad 3 \quad 8 \quad - \quad + \quad \times \quad =$$

Press any of them in any order, but use only four.

Make these numbers. Write which keys you pressed.

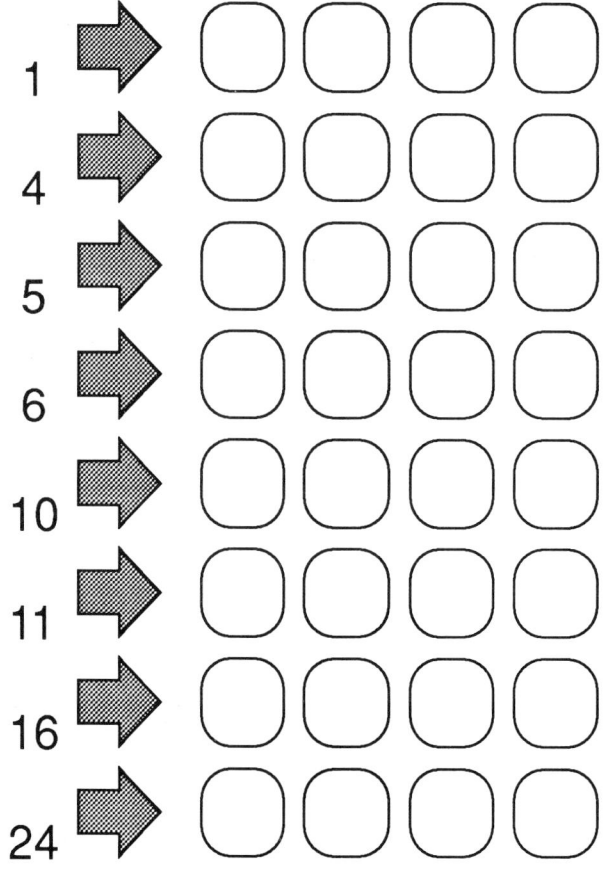

1 ◯◯◯◯

4 ◯◯◯◯

5 ◯◯◯◯

6 ◯◯◯◯

10 ◯◯◯◯

11 ◯◯◯◯

16 ◯◯◯◯

24 ◯◯◯◯

What other numbers could you make?

What if you used 1, 3 and 9, instead of 2, 3 and 8?

2

Tom counted his books in fours. He had three left over.

He counted them in fives. He had four left over.

How many books did Tom have?

Make up another problem like this.

Give it to a friend to solve.

1 You need 20 cubes.

Put the cubes in four piles.

The first pile must have four more than the second.

The second pile must have one less than the third.

The fourth pile must have twice as many as the second.

How many cubes are there in each pile?

2

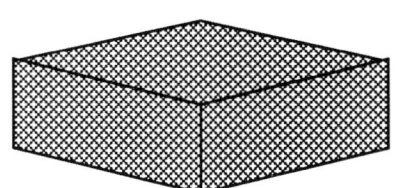

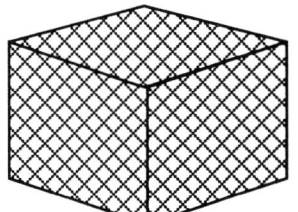

The red parcel balances
10 cubes.

The blue parcel balances
12 rubbers.

3 cubes balance 4 rubbers.   Which parcel is heavier?

3 Leroy and Sarah have the same number of pencils.

Leroy has three full boxes and four loose pencils.

Sarah has two full boxes and twelve loose pencils.

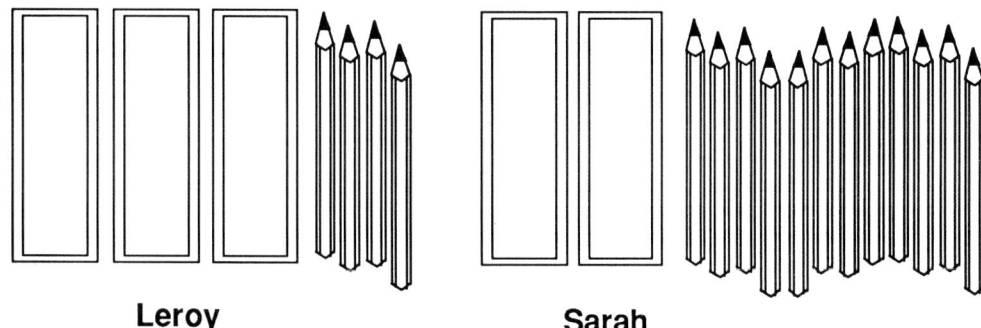

Leroy   **Sarah**

Each box holds the same number of pencils.

How many pencils are there in a box?

Make up a similar problem.   Ask a friend to solve it.

You need some plain paper to record your work and these number cards.

1 Investigate magic triangles. Put one number in each circle. The numbers along each side of the triangle must add up to the same total.

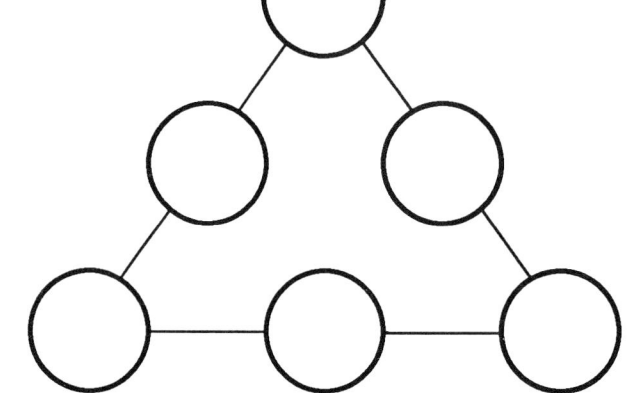

a. Use 1, 2, 3, 4, 5, 6.

b. Use 1, 2, 3, 5, 6, 7.

c. Use 1, 2, 3, 4, 6, 7.

Investigate different ways of doing it.

2 Use the numbers 1, 2, 3, 4, 5, 6. Put one number in each circle. Find the difference between each pair of linked circles. Add up all nine differences.

Arrange the numbers so the total of the differences is as big as possible.

1 These numbers are **consecutive**.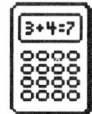
They follow one after the other.

2, 3, 4, 5, 6

The consecutive numbers from 2 to 6 add up to 20.

Make each number from 5 to 40 by adding consecutive numbers.

Keep a record.

5 = 2 + 3
6 = 1 + 2 + 3
. . . .
40 = 6 + 7 + 8 + 9 + 10

What patterns do you notice?

Can you make 1000 by adding consecutive numbers?

2 You need your calculator.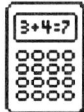
You may press only these keys.

Which numbers from 3 to 30 can you make?

Which is the largest number you cannot make?

What if you used only these keys?

④ ⑦ + =

Which is the largest number you cannot make?

3 Find different ways of making this sum correct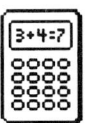
by putting plus or minus signs in suitable places.

1 2 3 4 5 6 7 8 9 = 100

1   Use these five number cards.

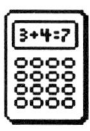

| 1 | 2 | 3 | 4 | 5 |

Arrange the cards to make these answers as large as possible.

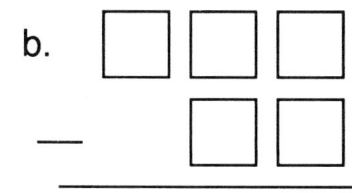

a.

b.

Arrange the cards to make these answers as small as possible.

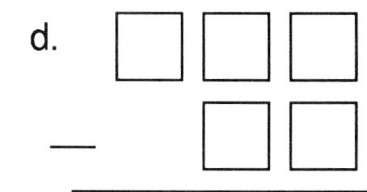

c.

d.

2   The numbers in the circles have been added in pairs.
The sum of each pair is in the box between the circles.
Complete the second diagram.

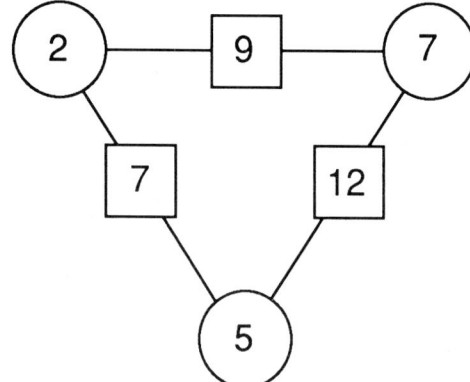

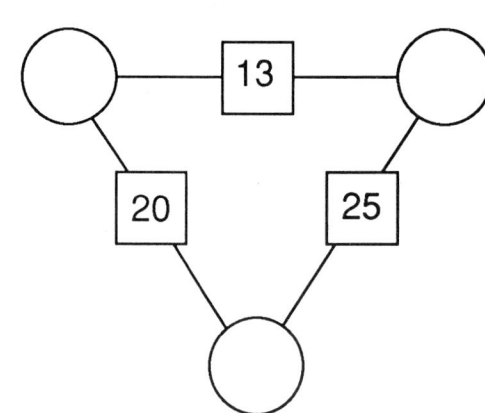

3   You can divide me exactly by 5. When my two digits are
reversed the number is 27 more than I am. Who am I?

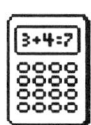

You need some squared paper.

1 Rashid is trying to complete a magic square.

Each row, column and diagonal must add up to the same number.

He needs to change over three pairs of numbers.

Finish the magic square for him.

13	7	12	4
3	10	5	15
2	11	8	14
16	6	9	1

2 The four numbers in the square on the left have been added, down, across and diagonally. The totals are shown in the circles. Complete the square on the right.

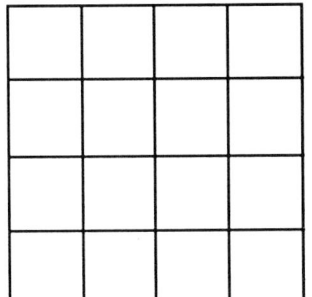

3 Put one digit in each square. This makes two numbers reading across and two reading down. Add up these four two-digit numbers.

2	4
1	2

24 + 12 + 21 + 42 = 99

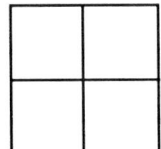

Make 67

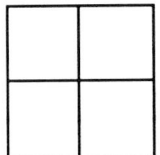

Make 111

1 Find the missing digits.

a.
$$\begin{array}{r} 4\ \square \\ +\ \square\ 8 \\ \hline 7\ 4 \end{array}$$

b.
$$\begin{array}{r} 3\ \square \\ -\ \square\ 9 \\ \hline 9 \end{array}$$

c.
$$\begin{array}{r} 3\ \square \\ +\ \square\ 7 \\ \hline 1\ 2\ 0 \end{array}$$

2 Complete these.

Put a number in each box. Use only 17, 28, 29, 37 or 43.

a. $\square - \square = 6$

b. $\square - \square = 8$

c. $\square + \square = 80$

d. $\square - \square = 20$

e. $\square - \square = 9$

f. $\square + \square = 45$

g. $\square + \square = 65$

h. $\square - \square = 11$

i. $\square - \square = 26$

j. $\square + \square = 71$

3 You need four 5cm rods and four 6cm rods (or strips of paper).
Using the rods, how can you measure these?

a. 1cm = 6cm − 5cm

b. 3cm

c. 9cm

d. 11cm

e. 13cm

f. 14cm

g. 16cm

h. 17cm

Investigate other lengths you could measure up to 40cm.

What if you used 5cm and 7cm rods?

1 A burglar has six bags of loot and two suitcases.

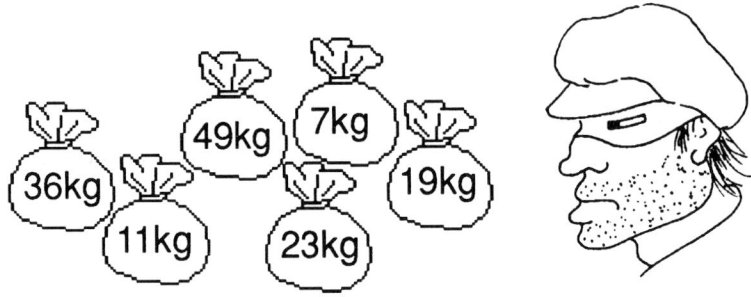

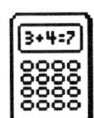

He wants to balance the load he must carry as best as he can.
How should he pack his two suitcases?

2 Each line in this magic star adds up to 28.

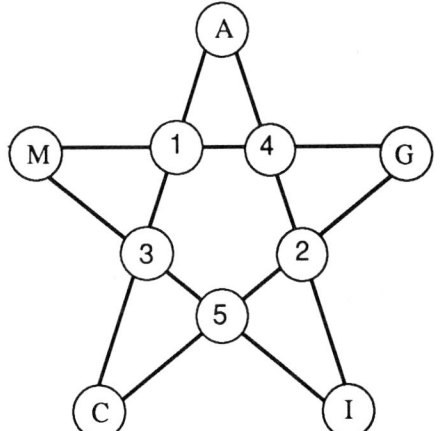

Find the numbers the letters
MAGIC stand for.

3 Use these numbers.

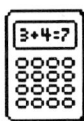

17   4   67   5   24   31   46   19

Put them in two groups of four.
Add the numbers in each group.
Find the difference between the two totals.
Like this, the difference is 27.

$$\boxed{24 + 31 + 46 + 19} - \boxed{17 + 4 + 67 + 5} = 27$$

Investigate the smallest difference you can make.

1 A girl, her brother and a dog stood in pairs on a weighing machine.

The girl and boy weighed 82kg.

The girl and the dog weighed 47kg.

The boy and the dog weighed 65kg.

How much did each one weigh?

2 Find the times the train is due at each stop.

9:00		9:15					
9:40						10:35	

10 minutes

London   Wimbledon   Woking   Basingstoke

How long is the journey from London to Wimbledon?

How long is it from Woking to Basingstoke?

The train waits at Basingstoke for 15 minutes before going back to London. Complete a timetable for the return journeys.

Basingstoke		
Woking		
Wimbledon		
London		

You need some squared paper and your calculator.

1 Use only these keys.

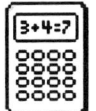

$$\left(1 \right) \left(0 \right) \left(5 \right) \left(+ \right) \left(= \right)$$

Make each of these numbers. Press as few keys as possible.

16, 37, 88, 638

Record how you did it.

2

275	382	81	174
206	117	414	262
483	173	239	138
331	230	325	170

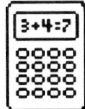

Choose any four numbers from the grid and add them up.

Find as many ways as possible of making 1000.

Design a 3 x 3 grid. Fill it with numbers so that as many as possible groups of three numbers add up to 1000.

3 Make 9.   Use only these keys, once each.

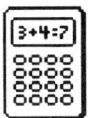

$$\left(1 \right) \left(2 \right) \left(3 \right) \left(4 \right) \left(5 \right) \left(6 \right) \left(7 \right) \left(8 \right) \left(9 \right) \left(+ \right) \left(- \right) \left(= \right)$$

Record how you did it.

1 The greengrocer had 32 apples and some empty baskets.

She put 9 apples in some baskets.

She put 2 apples in each of the other baskets.

How many baskets were there?

Make up another problem like this.

Ask your friends to solve it.

2 Use each of these six number cards.

Arrange the cards, one in each box, to make the answer correct.

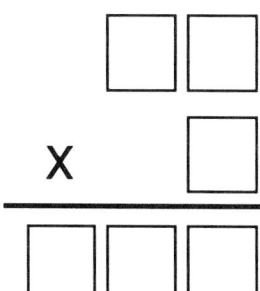

3 The area of a rectangle is 91 square centimetres.

What is its perimeter?

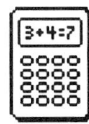

4 Using each of the digits 1, 2, 3 and 4 once only, and any of the signs $+$, $-$, x or $\div$, make each of the numbers from 1 to 40.

$$41 = 1 \times 43 - 2$$

$$42 = 41 - 2 + 3$$

Investigate other numbers you can make using 1, 2, 3 and 4.

You need some squared paper.

1 With 12 square tiles you can make just 3 different rectangles.

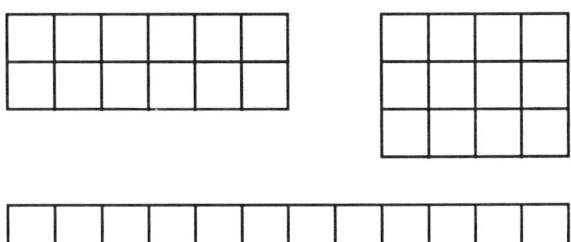

How many tiles do you need to make just 5 different rectangles?

2 Cross out just two numbers in this grid so that for each row and each column the sum of the numbers left is a multiple of 5.

1	2	4	8
5	3	2	3
7	7	1	6
2	6	3	9

Which numbers should you cross out?

3 You need a red pen and a yellow pen.

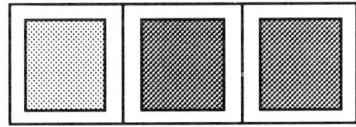

A man is fitting coloured glass in a window.

The window has three parts.

In each part he can put red glass or yellow glass.

How many different windows can he make?

How many different windows can he make if he has red glass, yellow glass and green glass?

What if he has four different colours?

1 Copy and complete this multiplication table.

x		4	9	
		8	18	
3		12		
	35			14
				2

2 Two consecutive numbers multiply to give 30.

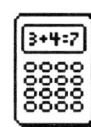

$$5 \times 6 = 30$$

Use your calculator to find two consecutive numbers which mutiply to give 182.   What are they?

Now find pairs of consecutive numbers to complete these.

a. ... x ... =   90   d. ... x ... =   756

b. ... x ... = 210   e. ... x ... = 1892

c. ... x ... = 342   f. ... x ... = 3306

3 Jake keeps rabbits and geese.
His animals have 40 heads and 88 feet between them.
How many rabbits does Jake have?
How many geese?

Make up another puzzle like this for your friends to do.

Maths workshop   © Cambridge University Press 1992

1 The number 18 is twice the sum of its digits.

$$18 = 2 \times (1 + 8)$$

The number 27 is three times the sum of its digits.

$$27 = 3 \times (2 + 7)$$

a. Which numbers are four times the sum of their two digits?

b. Which numbers are five times, six times, seven times, eight times or nine times the sum of their two digits?

c. What do numbers which are exact multiples of the sum of their digits have in common?

2 On a bike-riding holiday, Sue rode 105 miles in 5 days. Each day she rode 6 miles less than the previous day.

a. How many miles did she ride each day?

b. If she continued to ride 6 miles less each day, what was the greatest distance she could have ridden?

3  An octopus has two eyes and 8 arms. A mermaid has two eyes, 2 arms and a tail. A dolphin has two eyes and a tail.

A group of sea creatures has 32 eyes, 38 arms and 13 tails between them. How many of each kind are there?

1   Use each of these nine number cards.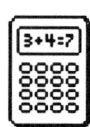

| 1 | 2 | 3 | 4 | 5 | 6 | 7 | 8 | 9 |

Arrange the cards in a 3 x 3 square so that:

the top number multiplied by 2 gives the second row;

the top number multiplied by 3 gives the third row.

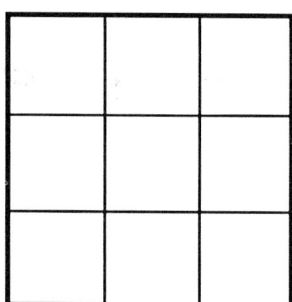

2   Use your calculator to find: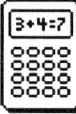

     a.   37 x 3   b.   37 x 6   c.   37 x 9

What do you notice about the answers?

Guess and write down the answers to these.

     d.   37 x 12   e.   37 x 15   f.   37 x 18
     g.   37 x 21   h.   37 x 24   i.   37 x 27

Check using your calculator.

Why do you think this happens?

3   Rose keeps cows and chickens.

Her animals have 70 eyes and 94 feet.

How many cows does Rose have?

How many chickens?

1 The digits of 322 are 3, 2 and 2.

The digits of 322 add to 7, since 3 + 2 + 2 = 7.

| 322 |

Investigate numbers whose digits add to 5.

You may not use the digit 0.

How many different numbers are there?

What is the difference between the smallest and the largest?

2 This sum is in code. Each letter stands for one of these digits:

1, 2, 5, 6, 7, 9

Crack the code! Find out which digit each letter stands for.

```
    L A Y
  + G A Y
  -------
  E G G S
```

3 Use these ten number cards.

| 0 | 1 | 2 | 3 | 4 | 5 | 6 | 7 | 8 | 9 |

Arrange the cards to make this answer correct. Use each card.

Find another way of making the correct answer using each card.

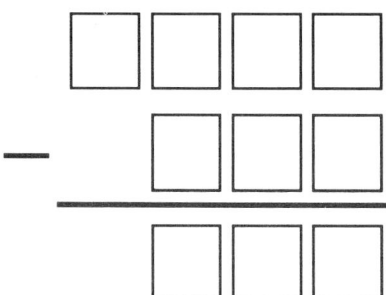

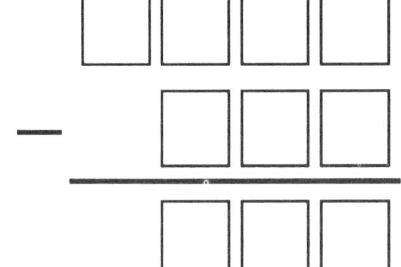

1 Each letter A to G is a code for one of these digits:

$$1, \ 3, \ 4, \ 5, \ 6, \ 8, \ 9$$

These calculations are all in the code.

A + A = B		A x A = DF
C + C = DB		C x C = BD
A + C = DE		A x C = EF
AA + CC = DAE		

What would be the code answer to DC x EF?

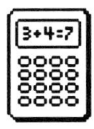

2 Some of the digits in these divisions are missing.
Each answer is a whole number. Find the missing digits.

⬜ 6 ÷ 7

8 ⬜ ÷ 9

5 ⬜ 6 ÷ 6

4 ⬜ 9 ÷ 7

4 ⬜ 7 0 7 ÷ 9

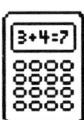

3 What numbers less than 100 have exactly three factors, including themselves and 1?

What number less than 100 has the largest number of factors?

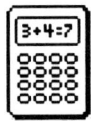

1

 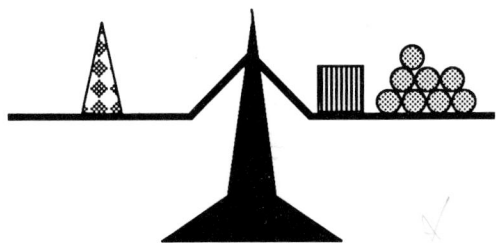

The scales balance like this.　　　And they balance like this.

How many spheres would balance the cone?

2

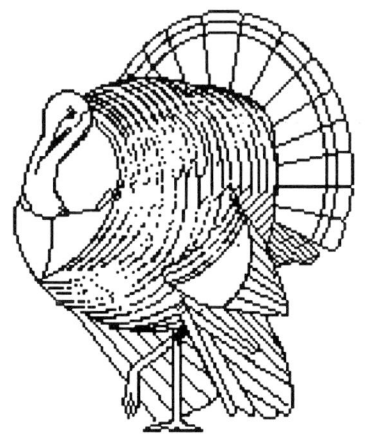

Farmer Tompkins has four turkeys.
She weighed them two at a time.

The weights of pairs of turkeys were
18kg, 20kg, 22kg, 26kg, 28kg and 30kg.

What is the weight of each turkey?

3　Imagine you have 27 balls and a balance, but no weights.
All the balls look alike, but one weighs more than the others.

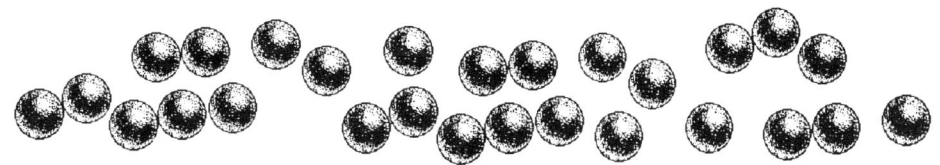

How can you find the heavier ball in just three balances?

1

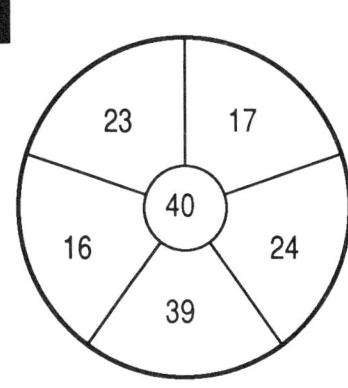

Karen, Tracey and Dean played darts.

Each threw six darts.

All the darts landed on the dartboard.

Only one dart hit the bullseye.

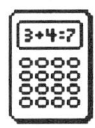

Karen scored 120, Tracey 110 and Dean 100.

What six hits did each of them make?

2 Find:

a. two consecutive numbers with a product of 3080;

.... X .... = 3080

b. three numbers with a product of 7511;

.... X .... X .... = 7511

c. three consecutive odd numbers with a product of 357 627.

.... X .... X .... = 357 627

3 5219 people voted for four candidates in an election.

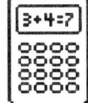

The winner got 22 votes more than one opponent, 30 more than another, and 73 more than the third.

How many votes did each of them get?

Winner	
2nd place	
3rd place	
4th place	

1 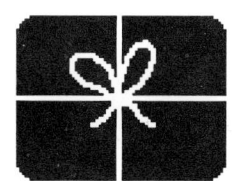 Anna posted some parcels.

She spent £6.60 on each one.

She used only 18p or 30p stamps, but each

parcel had a different combination of stamps.

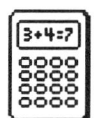

What was the greatest possible cost of all the parcels?

2 Use these nine number cards.

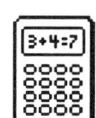

| 1 | 2 | 3 | 4 | 5 | 6 | 7 | 8 | 9 |

Use each card. Put one in each box to make the answer correct:

a. ☐☐☐☐ X ☐ = ☐☐☐☐

b. ☐☐☐ X ☐☐ = ☐☐☐☐

c. ☐☐☐ X ☐☐ = ☐☐ X ☐☐

Investigate different ways of doing it.

Find the largest product in each case.

3 Find the missing digits without using the ÷ key.

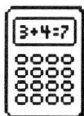

a. (2)( )(5)(4)( ) x (3) = ( )(2)( )( )(4)

b. (1)( )(2) x (1)(4)( ) = (2)(4)( )(4)(0)

1 It is said that all mammals, except humans, have the same number of heart beats during their lives.

The average life span of a cat is 15 years.

A cat's heart beats on average 120 times each minute.

What would be the average life span of these animals?

Deer:	90	beats per minute
Elk:	60	beats per minute
Dormouse:	450	beats per minute
Chamois:	72	beats per minute
Porcupine:	120	beats per minute

2 Find the missing digits.

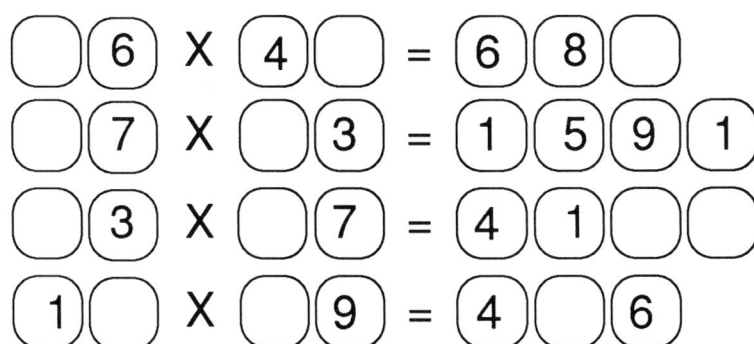

3 You need about 20 gallons of water for a bath.

There is an old saying: *A pint of water weighs a pound and a quarter.*

There are 8 pints in a gallon, and about 2.2 pounds to the kilogram.

Approximately how many kg of water are used for a bath?

1 You need a stop watch and a measuring tape.
What is the average walking speed of your friends?

Measure out a suitable distance.
Ask some friends to walk the distance at their
normal walking speed.

Use the stop watch to find the time taken by each friend.
Find the average (mean) time taken.

What is the average walking speed:

a. in metres per second?

b. in metres per minute?

c. in metres per hour?

d. in kilometres per hour?

Eight kilometres is about the same distance as five miles.
What is their average walking speed in miles per hour?

2 Jenny and Lisa rode their bikes.
Their route had three equal sections each four miles long.
Lisa went at a constant speed the whole way.
Jenny did the first section at twice Lisa's speed, the second section
at the same speed as Lisa, and the third section at half Lisa's speed.

Who finished first? How far ahead was she?

1   Use these five number cards.

| 1 | 2 | 3 | 4 | 5 |

Put one card in each box. Make   Put one card in each box. Make
the answer as **big** as possible.   the answer as **small** as possible.

2   Use only these five keys on your calculator.

(1)   (0)   (+)   (=)   (·)

Make each of these numbers. Press as few keys as possible.

0.12,   2.4,   0.88,   1.04,   2.21

Record how you did it.

3   What number, when multiplied by itself, gives the answer 10?

.... X .... = 10

Draw a diagram to show how you solved this problem.

4   The sum of two numbers is 10. Their product is 19.71.
What are the numbers?

1 Refugees leave their own countries for fear of being persecuted.

The table shows the number of refugees in some different countries.

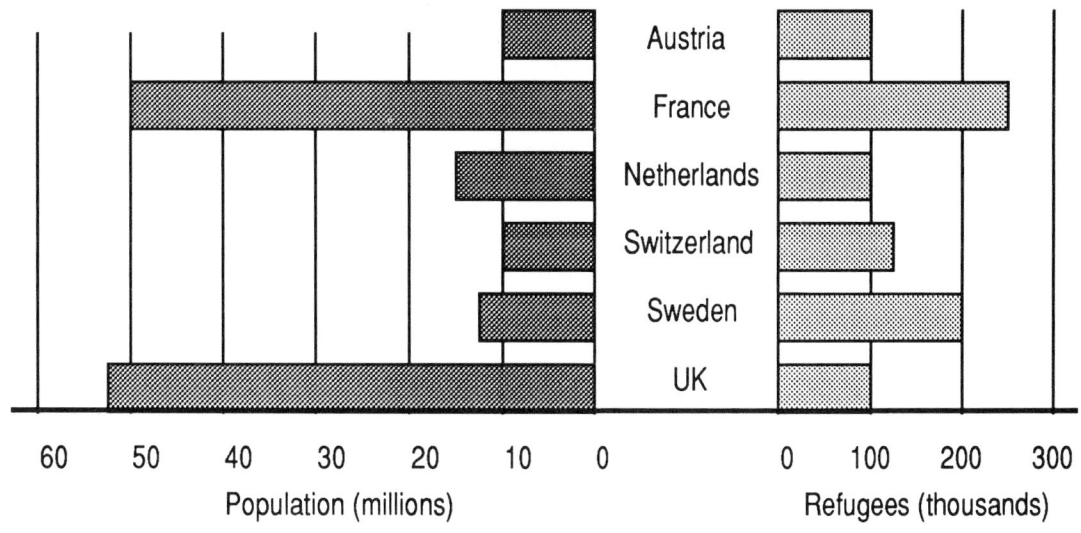

| 60 | 50 | 40 | 30 | 20 | 10 | 0 | | 0 | 100 | 200 | 300 |

Population (millions)   Refugees (thousands)

As a fraction of its population:

    a. which country has most refugees;

    b. which has the fewest?

2
Fry D.G, 46 Edward Av......................Eastleigh 643561   Fry W.E, Rosebower,Stratford Sub Castle......   Salisb
Fry D.G, 14 Laurel Rd   Locks Hth 2521   Sotc
Fry D.G, 35 St Marys   Eastleic
Fry D.J, 15 Spring Rd   .. Sotc
Fry D.M, 11 Haywain   You need a telephone directory.   Roms
Fry D.M, 14 Meon Gd   Sotc
Fry D.M, 30,Raglan C   ....Sotc
Fry D.P, 23 Castle Rd   ....Soton 453262
Fry E, 7 Warwick Ct,N   ....Soton 770580
Fry E.C, 50 Kingsfield   Estimate the total number of entries.   ....Soton 775379
Fry E.C, 66 Wakefield   Cholderton 213
Fry E.E, 4 St Davids   ..Andover 65783
Fry E.E, 98 Foundry I   .....Soton 556165
Fry E.J, 64 Victoria R   . . . .   Winchester 69357
Fry E.J, 7 Woodstock   ... Soton 462133
Fry E.L, 3 Bankside,V   .. Soton 731170
Fry E.M, 25 Brookside   Describe how you did it.   .... Soton 445242
Fry E.W, 31 Sedbergh   . Locks Hth 82797
Fry F, Nestyn,Middle   . Salisbury 22090
Fry F.G, 15 Newbridge   .. Hythe 845648
Fry G –   ...Soton 866492
    (Res) Wolfhanger   Salisbury 743659
Fry G, 1 The Bungalows,Enford .... Stonehenge 70929   Fryer E.W.M, Annexe 7,Old Inn Dv,Townhill Pk. ......Soton 552297

3 a. How many great great great grandparents do you have?

   b. On average, each generation is
          25 years older than the next.
          Estimate how many great ... great
          grandparents you had 1000 years ago.

Find out what the population of the world was then.

1 Find the missing digits.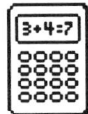

⑨①②⑧③⑦◯   is divisible by 8.

⑧◯②③⑤①   is divisible by 9.

◯⓪③④⑧②   is divisible by 11.

②⑦◯③⑥◯⑤   is divisible by 225.

2 7896 is a remarkable number.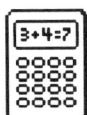

78	is divisible by 2.
789	is divisible by 3.
7896	is divisible by 4.

Arrange the digits 1, 2, 3, 4, 5, 6 to make a remarkable number so:

> the first two digits are divisible by 2,
>
> the first three digits are divisible by 3,
>
> . . . and so on until finally
>
> the first six digits are divisible by 6.

Find different ways of doing it.

Make a ten-digit remarkable number using each digit 0 to 9.

3 17 ÷ 7 is 2, remainder 3, but a calculator shows 2.4285714.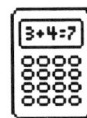
Investigate. Use your calculator to divide one number by another and find the remainder as a whole number.

If I divide one whole number less than 20 by another,
my calculator shows 1.1818181.
What were the two numbers?

What if the calculator showed 0.5384615?

4 Find a number which has exactly 15 different divisors.

1 You have one of each of these coins: 10p, 5p, 2p and 1p.

How many different amounts can you pay?
Write them in your book.

2 You have silver coins only. You can use as many as you wish.
How many different ways can you find of making 50p?

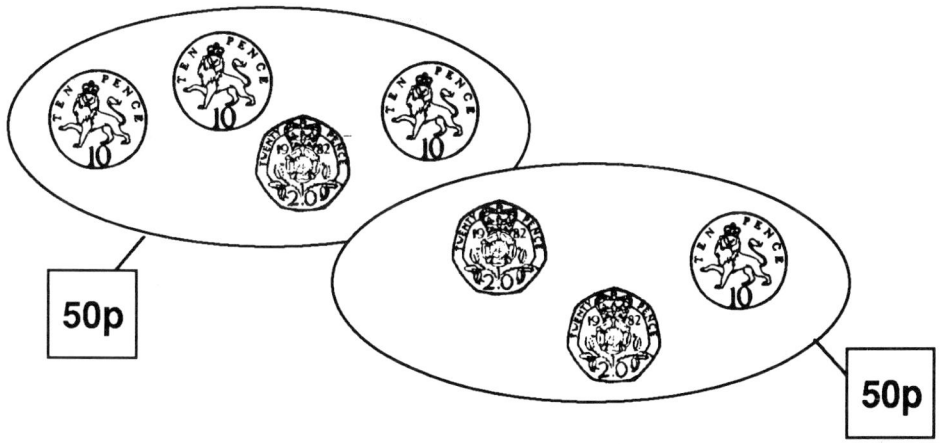

50p

50p

Record each way in your book.

3 A mother had £1.95 in silver coins.
She divided them between her 3 children.

They each got the same four coins.

Which coins were they?

1 What is the least number of coins you need for each of these?

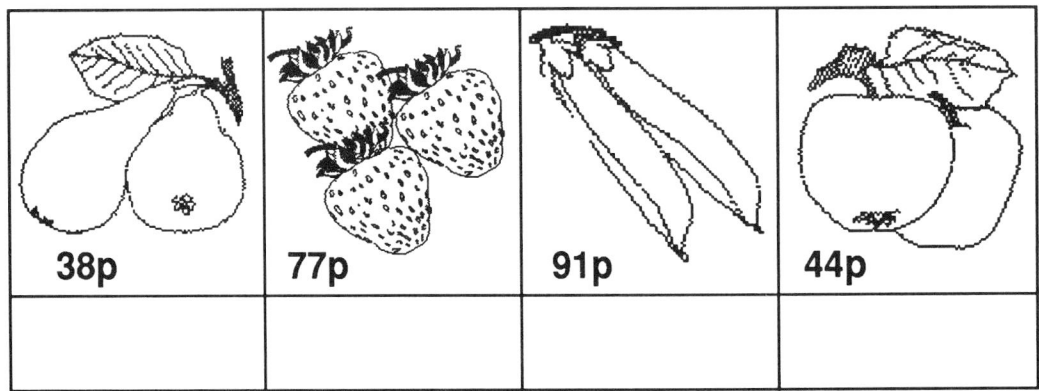

38p	77p	91p	44p

2 You have two 15p stamps and two 20p stamps.

You can send a parcel without buying more stamps.

What might it cost?

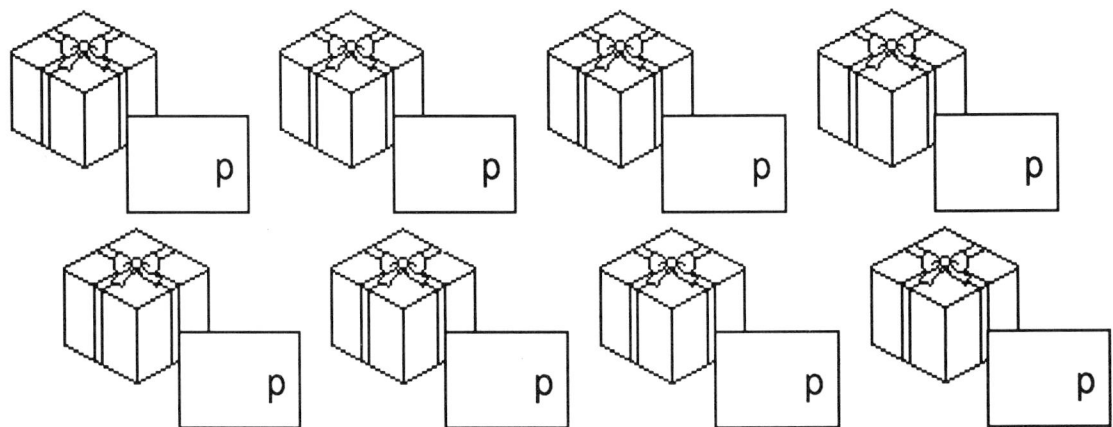

3 Mrs Singh spent £2 on 10p and 20p stamps.
She bought three times as many
10p stamps as 20p stamps.

How many of the stamps did she buy?

1 You need one of each of these coins.

You have one of each of these coins: 50p, 20p, 10p, 5p, 2p and 1p.

How many different amounts can you pay?

2 You have three 15p stamps and three 20p stamps.

You can send a parcel without buying more stamps.

Investigate what it might cost.

How many different possibilities are there?

3 The sign on the bus says: *Please give the exact fare.*

What is the largest sum of money you could have in coins without being able to pay exactly £1 for a ticket?

4 Maria spent £1 on stamps.

She bought some 4p stamps, six times as many 2p stamps, and spent the rest on 5p stamps.

How many of each kind of stamp did she buy?

1 In this bargain offer, the average cost of an apple is reduced by 5p.

BARGAIN

Spend £2 and get 2 free apples

How many apples did I get for £2?

2

What amounts up to £1 can you only pay exactly by using at least three coins?

What if you had to use at least four coins?

3 You need some squared paper.

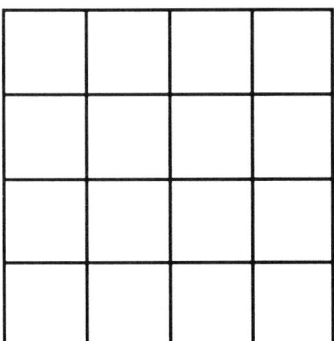

A card for stamps has 16 spaces.

You have plenty of 1p, 2p, 3p, 4p and 5p stamps.
Stamps of the same value must not be placed in the same straight line, horizontally, vertically or diagonally.

What is the greatest value you can stick on the card?

1 Four footballs and three tennis balls cost £15.60.
Three footballs and four tennis balls cost £15.20.

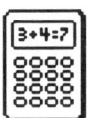

a. What does a tennis ball cost?

b. What does a football cost?

2

A factory makes teddy bears in five sizes.

Size 1:	£1.80
Size 2:	£2.70
Size 3:	£3.40
Size 4:	£3.90
Size 5:	£4.80

A playgroup spent up to £10 on three different sized bears.
Investigate which three bears the playgroup bought.
How many different possibilities are there?

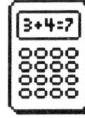

3 Lisa, Jacky and Pat bought some balloons.

Each spent exactly 60p in different ways.
Each one bought some 3p balloons.
Each one bought some 5p balloons.
Pat bought twice as many 5p balloons as Lisa.

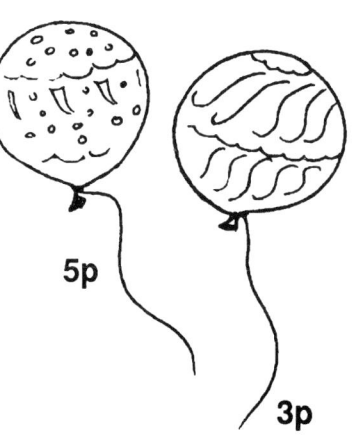

5p

3p

a. Who bought the most balloons?

b. Who bought the fewest balloons?

1 Mrs Choy spent £10 on 100 eggs for her shop.

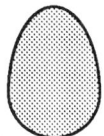 **Large eggs**
50p each

 **Medium eggs**
10p each

 Small eggs
5p each

For two of the three sizes, she bought the same number of eggs.

How many of each size did she buy?

2 Ann, Mary, Jane and Kate each have a brother.

The sisters and brothers went shopping with £32 between them.

They spent all their money.

Ann spent £1.	Andrew spent the same amount as his sister.
Mary spent £2.	Tim spent twice as much as his sister.
Jane spent £3.	Nick spent three times as much as his sister.
Kate spent £4.	Simon spent four times as much as his sister.

What was the name of each girl's brother?

3 You can go from Andover to Cambridge on the train, via London.

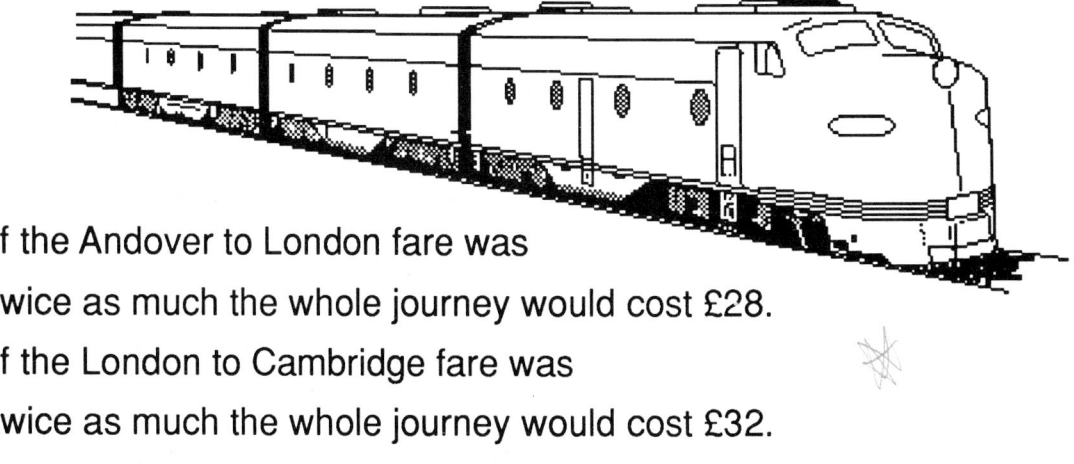

If the Andover to London fare was

twice as much the whole journey would cost £28.

If the London to Cambridge fare was

twice as much the whole journey would cost £32.

What is the rail fare from Andover to Cambridge?

1 The golf course at Uphampton has 18 holes.

A good golfer would take 72 strokes (shots) for a round of 18 holes.

If a player takes 70 strokes, this is a score of -2, or 2 under.

If a player takes 78 strokes, this is a score of 6, or 6 over.

After three rounds of a tournament, these were the players' scores.

Name	Score	Name	Score
Andrews	-9	Fielder	0
Brown	-8	Green	1
Cox	-8	Howard	1
Dobson	-4	Innes	4
Evans	0	Jones	6

The numbers of strokes taken on the fourth round were:

Brown	67	Evans	71
Cox	69	Dobson	71
Andrews	70	Howard	72
Green	70	Fielder	74
Jones	70	Innes	74

Make a table showing, for each player:

a. the final score;

b. the total number of strokes taken on all four rounds.

2 Dobson and Howard played against each other for one round.

Their scores on each of the 18 holes were as follows.

Dobson	1	0	-2	3	0	-2	-1	0	1	1	-2	4	-1	0	1	-1	0	-2
Howard	0	1	-1	2	-1	-1	1	0	-1	2	-1	-1	1	0	2	-1	-1	1

How many strokes did each of them take for that round?

Which player won the most holes?

1 You need some squared paper and some coloured pens.

Draw a 3 x 4 rectangle of 12 squares.

Colour in $\frac{1}{2}$, $\frac{1}{4}$, $\frac{1}{6}$ and $\frac{1}{12}$ of the rectangle.

Use a different colour for each fraction. Colours must not overlap.

You should have coloured the whole shape.

Now draw a 4 x 5 rectangle. Colour it so that it is filled with different fractions each of which has 1 as the top part.

Write the fractions below the rectangle.

Now try a 5 x 6 rectangle.

Investigate other rectangles which you can divide and fill with different fractions each of which has 1 as the top part.

2 This problem is based on one written 1400 years ago by the great Indian mathematician Brahmagupta.

Brahma gave out all the sweets in a bag.

He gave one third to Rupa.

He gave one quarter to Bashir.

He gave one fifth to Gita.

He kept 13 for himself.

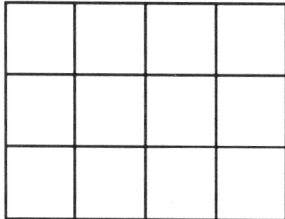

How many sweets were there in the bag?

3 Find a fraction bigger than one quarter and less than one third.

1 Is $\frac{51}{101}$ more or less than 0.5?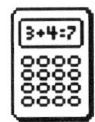

$$\frac{50}{100} = 0.5 \qquad \frac{51}{101} = ?$$

Investigate adding 1 to the numerator and denominator of a fraction.

Does it get bigger or smaller?

Try it on these to start with: $\frac{4}{5}$, $\frac{1}{200}$, $\frac{17}{9}$. What happens?

Try some other fractions.

What happens if you subtract 1?

2 a. If you add 1 to my numerator, my value is one third.

   If you add 1 to my denominator, my value is one quarter.

   Who am I?

b. I am a number less than 20.

   Two less than half of me is one more than one third of me.

   Who am I?

3 Join the broken strings to the number line so that the fraction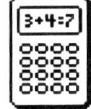
balloons are tied to the line in the correct order.

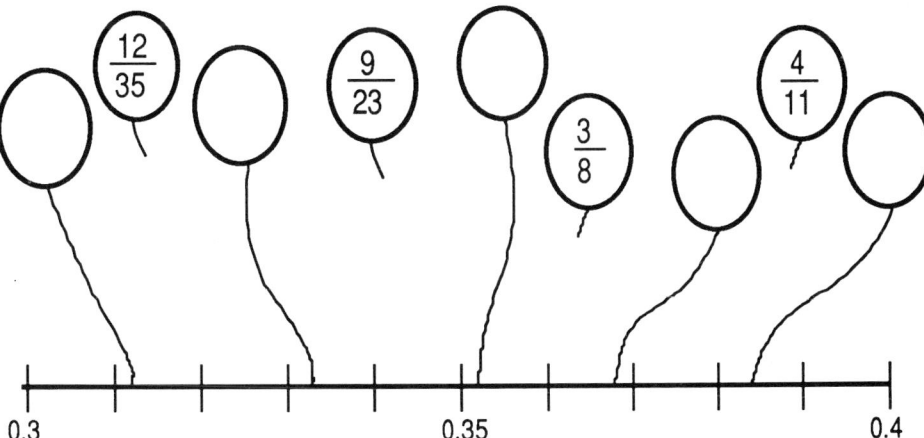

Put fractions in the empty balloons so they are all tied in order.

4 Find a fraction greater than $\frac{7}{17}$ but less than $\frac{13}{31}$.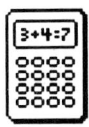

1 I am a number less than 20.

Two less than half of me is one more than a third of me.

Who am I?

2 Parveen and Afzal are two teenagers.

The difference between one-sixth of Parveen's age and one-seventh of Afzal's age is one year.

How old is Parveen?   How old is Azfal?

3 You can travel through this maze either horizontaly or vertically.

Cells in the maze can be visited only once.

As you pass through a number, add it to your score.

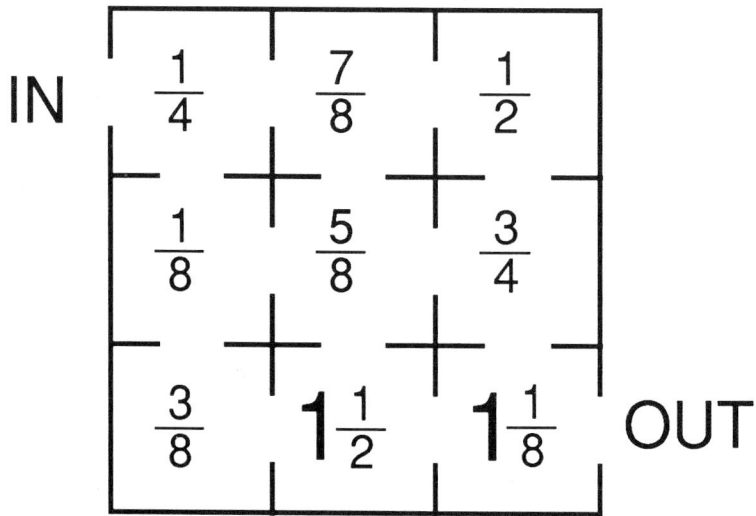

Which path gives a total of $4\frac{3}{8}$?

What is the least total for a path?

What is the greatest total for a path?

Investigate the greatest and least totals when you alternately add and subtract numbers.

1 Two fractions, each with numerator 1, are turned into decimals.
The calculator shows:

$$0.0204081$$

and $$0.0114942$$

What were the two fractions?

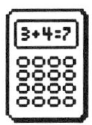

2

$\frac{1}{7}$	
$\frac{2}{7}$	
$\frac{3}{7}$	
$\frac{4}{7}$	
$\frac{5}{7}$	
$\frac{6}{7}$	

Use your calculator to work these out as decimals.

What do you notice about the first six digits after the decimal point?

Without your calculator, write the decimal fractions for $\frac{8}{7}$, $\frac{9}{7}$ and $\frac{16}{7}$.

If your calculator displayed more digits, what would be the first twelve digits after the decimal point for $\frac{1}{7}$?

3 Enter a number in your calculator, say 9. Divide by 5. Add 1.

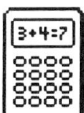

$\boxed{9} \; \boxed{÷} \; \boxed{5} \; \boxed{+} \; \boxed{1} \; \boxed{=}$ Did you get 2.8?

Now start with 2.8, $\boxed{÷} \; \boxed{5} \; \boxed{+} \; \boxed{1} \; \boxed{=}$ Did you get 1.56?

Carry on the chain, and keep a record of what you get each time. What do you notice?

Start with a different number, say 7.   What happens this time?

Try starting with some other numbers. Can you find a rule?

1      Lena's dad wanted to buy a second hand car.
The garage said he could have a 20% discount but
he must also pay 15% for a six month's guarantee.

Should Lena's dad:

a.   claim the discount first and then pay the guarantee;

b.   pay the guarantee first and then claim the discount?

2   A farmer bought two cows. Later, he sold them for £600 each.

   He made a loss of 20% on one cow,
and a profit of 20% on the other cow.

Did he make a profit or a loss on the whole deal?

3   A company director was given a pay rise.
In the first year, she was given a 50% pay rise.
In the second year she had a 20% pay rise.
A newspaper reported her pay rise as 70% over two years.

Was the newspaper right?

4      Julie carried out a survey.
She asked people if they liked lemonade.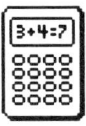

She used her calculator.
63.41463% of those asked said yes.

There were fewer than 50 people in Julie's survey.
How many of them liked lemonade?

1 Pair off the numbers in the ring.

Each pair must add to the same number.

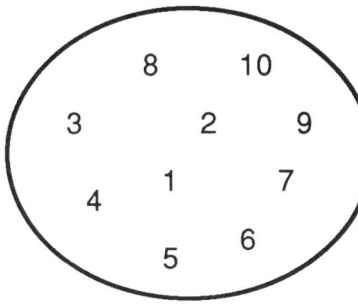

2 Now try these.

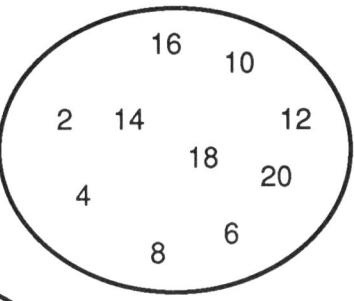

3 Draw some rings.

Put numbers in them.

Each pair must add

to the same number.

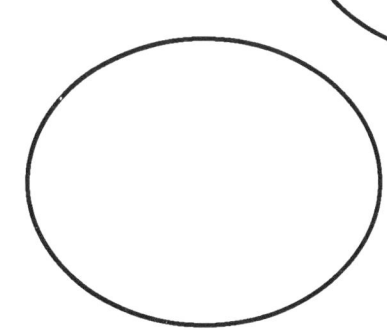

4 You need some squared paper.

The staircase on the right has 5 steps.

It is made from 15 squares.

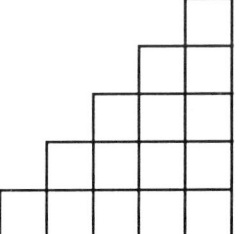

This staircase has 3 steps.

It is made from 6 squares.

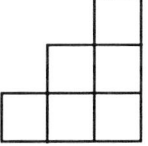

Investigate other staircases. Fill in the staircase chart.

Number of steps	1	2	3	4	5	6	7	8	9	10
Number of squares			6		15					

How many squares would there be
in a staircase with 20 steps?

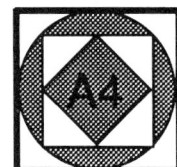
You need some squared paper.

1 Copy and complete these patterns.

☐ ☐ 12, 15, 18, 21, ☐ ☐

☐ ☐ 16, 20, 24, 28, ☐ ☐

☐ ☐ 18, 16, 14, 12, ☐ ☐

Make up a pattern of your own like this.

☐ ☐ . . , . . , . . , . . , ☐ ☐

2 Copy these grids. Colour the pattern of 3 on each of them.

1	2	3	4	5
6	7	8	9	10
11	12	13	14	15
16	17	18	19	20
21	22	23	24	25
26	27	28	29	30

This grid has 5 columns

1	2	3	4	5	6	7
8	9	10	11	12	13	14
15	16	17	18	19	20	21
22	23	24	25	26	27	28
29	30	31	32	33	34	35
36	37	38	39	40	41	42

This grid has 7 columns

Would 46 be in the pattern of threes?
How do you know?

On squared paper, draw two grids with different numbers of columns.
The pattern of threes must go straight up and down. Colour it.

Draw a different grid. It must make a similar pattern of threes
as the grid with 5 columns.

Investigate the pattern of fours on different grids.

You need some counters and some squared paper.

1 Make the next few patterns in this sequence.

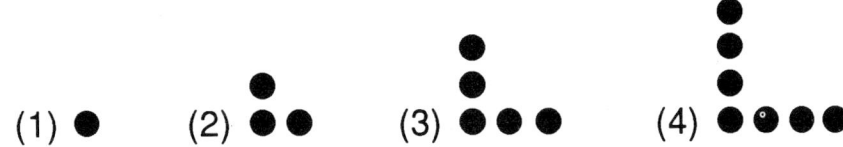

Copy and complete the table.

Position	1	2	3	4	5	6	7	8	9	10
Number of counters	1	3	5							

How many counters would there be in the 20th number?

How many in the 1000th number?

2 Now investigate this sequence.

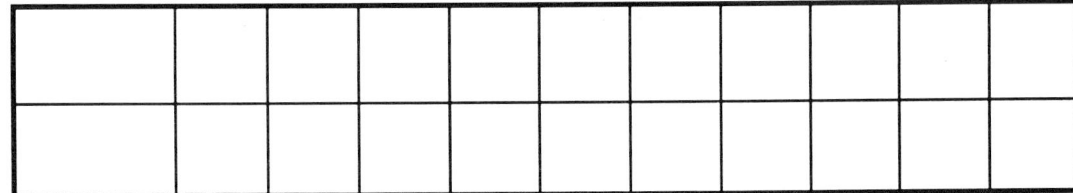

Make a table.

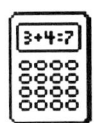

How many counters in the 20th number?

How many in the 100th number?

3 What is the smallest number that will divide exactly by 5, 7, 9 and 21?

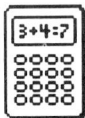

You need squared paper and a ruler.

1 Play **Pilots** with a friend.

Rules

- Mark a rectangle. Start is at the top right hand corner. Home is at the bottom left corner.
- Take turns to draw a straight line. It can be any number of steps long, but only in a S, SW or W direction.
- Each line must start where the last line ended.
- The winner is the first to draw a straight line to reach home.

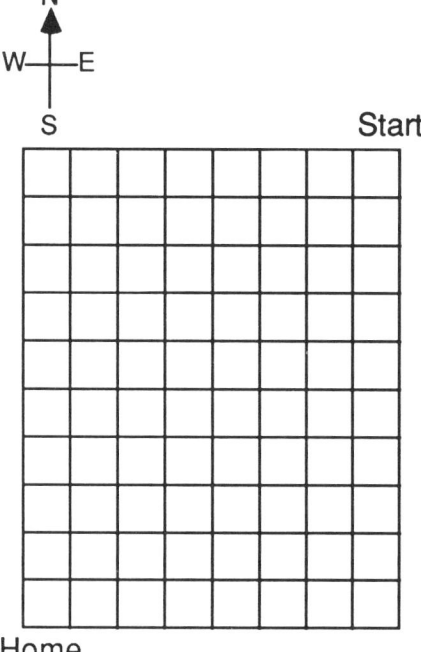

Investigate for different sized rectangles. Which are safe positions? Make up variations of the game.

2 A ball starts from A in the top left corner of a rectangular table. It bounces across diagonally.

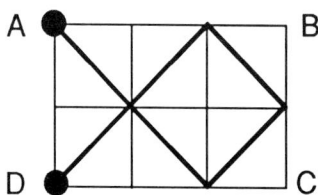

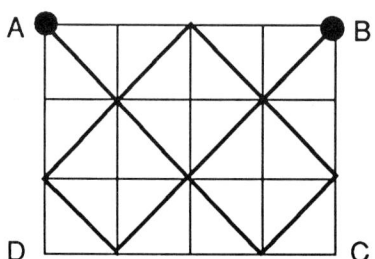

Investigate for different rectangles. Make a table of your results.

AB	3	4							
AD	2	3							
End	D	B							

Can you predict in which corner the ball will finish?

You need some counters and some squared paper.

1 Make the next few patterns in this sequence.

(1) 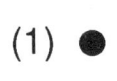   (2)    (3)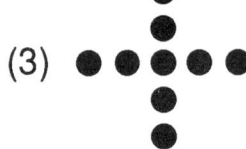

Copy and complete the table.

Position	1	2	3	4	5	6	7	8	9	10
Number of counters										

How many counters would there be in the 20th number?

How many in the 100th number?

2 Now investigate this sequence.

(1)    (2)    (3)   (4)

Make a table.

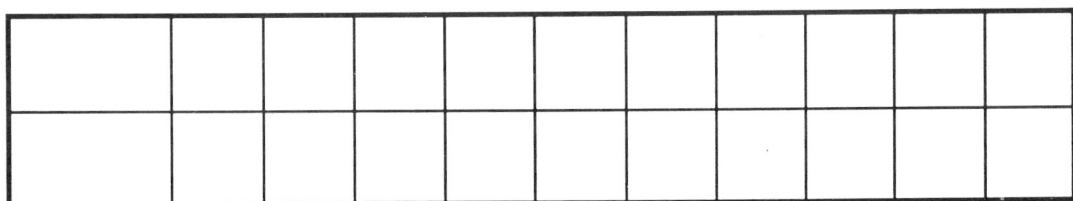

How many counters in the 20th number?

How many in the 100th number?

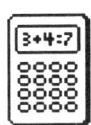

3

There are 7 people in the O'Grady family.

Everyone gives a present to each of the others.

How many presents are needed?

How many would be needed for

a family of 16 people?

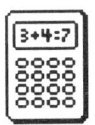

1 You need a paper strip about 50cm long. Fold it in half.

fold left over right

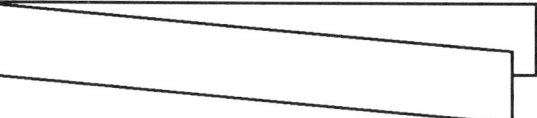

There is now one fold line
and two regions.

fold again left over right

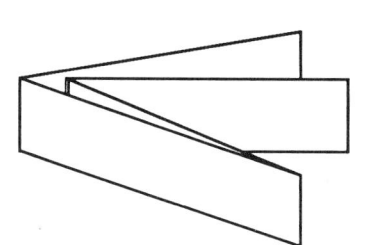

How many fold lines?
How many regions?

Keep folding in half. Record your results in a table.

Number of times folded	1	2	3	4	5	6	7
Number of fold lines	1						
Number of regions	2						

If you folded ten times, how many fold lines would there be?
How many regions?

2 In this puzzle, letters stand for numbers; a = 2, b = 3 and c = 5,
but you will need to work out what d, e, f and g stand for.

Across

1   b x d
3   a^5
6   13 x g
7   a^2 x b^2
8   48 x b^2
11 d^2
13 a x c x f
14 e^2
15 a^3 x d

Down

2   f^2
3   b x g
4   $a^4 + d$
5   $d^2 - b$
6   a x b^2
9   $(b + f)^2$
10 c x e
11 a x b x e
12 e x f
13 a x d

1	2	3	4	5
6			7	
	8		9	10
11	12	13		
	14		15	

1 Last year my age was a square number.

Next year it will be a cube number. How old am I?

How long must I wait before my age is both a square number and a cube number?

2 A 10cm x 10cm x 10cm cube has a capacity of 1 litre. What sized cube will hold half a litre?

Draw a diagram to show how you solved this problem.

3 Some numbers are equal to the sum of two squares. For example,

$$34 = 3^2 + 5^2$$

Which numbers less than 100 are equal to the sum of two squares?

Copy and complete the table.

+	1^2	2^2	3^2	4^2	5^2	6^2	7^2
1^2							
2^2							
3^2							
4^2							
5^2		34					
6^2							
7^2							

Investigate numbers less than 100 which are the sum of 3 squares.

4 5 appears as the last digit of its square: 25.

90 625 appears as the last digits of its square: 8 212 890 625.

$$90\ 625 \times 90\ 625 = 8\ 212\ 890\ 625$$

Investigate numbers less than 1000 which appear as the last digits of their squares. What about cubes?

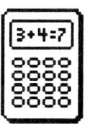

You need centimetre cubes and a long sheet of squared paper.

1

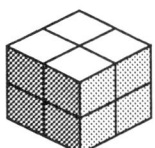

Build a 2cm solid cube from
the centimetre cubes.
How many do you need?

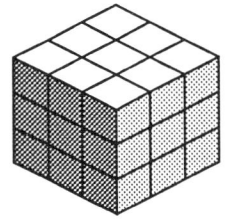

Build a 3cm solid cube.
How many do you need this time?

Copy and complete the table.

Edge of solid cube (cm)	1	2	3	4	5	6	7	8
Number of cm cubes	1	8						

Draw a graph to show how solid cubes grow.

How many cubes would be needed for a 20cm solid cube?

2 Now make some hollow cubes. Copy and complete the table.

Edge of hollow cube (cm)	2	3	4	5	6	7	8
Number of cm cubes	8	26					

Plot the figures for hollow cubes on the same graph as solid cubes.

What do you notice?

How many cubes would be needed for a 20cm hollow cube?

How many for an n-cm hollow cube?

Write it as a formula.

3 Investigate skeleton cubes.

Here is a 5cm skeleton cube.

Make a table.

Add the figures to the same graph.

Find a formula for skeleton cubes.

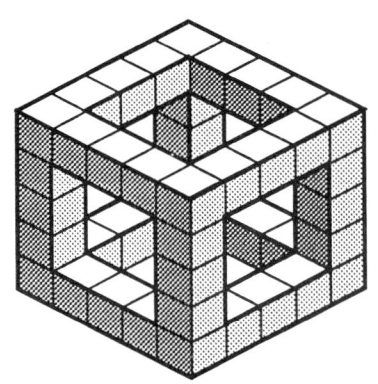

1 Imagine a pattern of counters in a long line.

The pattern starts like this: two red, four blue, two red, four blue, . .

What colour would the 65th counter be?

What position in the line would the 17th blue counter be?

This pattern is five blue, four red, five blue, four red, ...

What colour would the 65th counter be?

What position in the line would the 15th red counter be?

Make up some more puzzles like this. Ask your friends to do them.

2 Split the number 12. Multiply the parts together.

$$12 = 9 + 3 \qquad 9 \times 3 = 27$$
$$12 = 7 + 3 + 2 \qquad 7 \times 3 \times 2 = 42$$
$$12 = 6 + 3 + 2 + 1 \qquad 6 \times 3 \times 2 \times 1 = 36$$

Can you make a product of 32? How many ways can you do this?

Can you make 40? What about 44?

What is the largest product you can make?

Try starting with different numbers.

Investigate the largest product you can make.

1 Choose a two digit number.

Reverse it.

Find the sum.

```
   7 4
+  4 7
-------
 1 2 1
```
It is a multiple of 11.

Try a four digit number.

```
   2 8 6 1
+  1 6 8 2
-----------
   4 5 4 3
```
Is 4543 a multiple of 11?

Investigate other numbers. Does it always work?

2 A **palindrome** reads the same forwards or backwards.

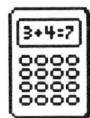

Take a number.

Reverse it.

Find the sum.

```
   2 4        3 6 2 4          8 6 2
+  4 2      + 4 2 6 3        + 2 6 8
------      ----------       -------
   6 6        7 8 8 7        1 1 3 0
                           +   3 1 1
                           ---------
                             1 4 4 1
```

The answers to these sums are palindromes.

Investigate other numbers.

How many additions are needed before you get to a palindrome?

3 Take any three digit number: for example, 875.

Reverse the digits to make another number.

Subtract the smaller from the larger.

```
   8 7 5
-  5 7 8
--------
   2 9 7
```

Take the answer.

Reverse the digits to make another number.

Add these two numbers.

```
   2 9 7
+  7 9 2
--------
 1 0 8 9
```

Investigate other three digit numbers. What happens?

1 In numbering the pages of a book,
555 digits were used.

How many pages were there?

How many 5s were used?

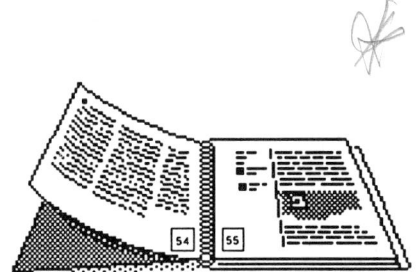

2 Choose any four different digits, except 0.
Using all four digits, make the largest and smallest
numbers possible. Find their difference.

For example, with 3, 6, 2 and 8 you would do this.

$$
\begin{array}{r}
8\,6\,3\,2 \\
-\ 2\,3\,6\,8 \\
\hline
6\,2\,6\,4
\end{array}
$$

Now use the four digits of the answer to make the largest and
smallest possible numbers. Find their difference.

$$
\begin{array}{r}
6\,6\,4\,2 \\
-\ 2\,4\,6\,6 \\
\hline
4\,1\,7\,6
\end{array}
$$

Keep repeating the process using the four digits in the answer as
the new starting point. What happens?

Investigate other sets of four digits.

What do you notice?

What is the longest chain of subtractions you can find?

3 There are 348 children at Church Street School.
There are 231 children under 9 years old, and 165 over 7 years old.
How many are of the children are 8 years old?

1 The symbols ◆ and ✳ are rules for combining numbers.

a. What does ◆ stand for?

5 ◆ 4 = 5

6 ◆ 9 = 9

6 ◆ 3 = 6

6 ◆ 5 = 6

b. What does ✳ stand for?

5 ✳ 2 = 12

3 ✳ 4 = 10

2 ✳ 5 = 9

4 ✳ 0 = 8

c. Make up some rules for combining numbers.

Write your rule. Give four examples.

2 The symbol ▲ is a rule for combining numbers.

2 ▲ 3 = 8

3 ▲ 2 = 9

3 ▲ 4 = 81

Write answers to these.

a. 4 ▲ 3

b. 2 ▲ 5

c. 5 ▲ 1

d. 5 ▲ 5

3 Copy and continue these two patterns.

1 ■ 4 ▼ 2 ■ 7 ▼ 5 ■ 16 ▼ 14 . . . .

18 ■ 9 ▼ 26 ■ 13 ▼ 38 ■ 19 ▼ 56 . . . .

4 Choose two numbers. Make a number chain by using the rule: *add the previous two numbers and write down the units digit.*

This is what happens if you start with 8 and 4.

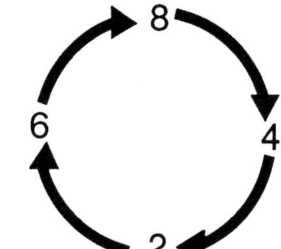

Investigate other pairs of starting numbers.

What is the shortest chain?

What is the longest chain?

What if you multiply, instead of adding?

What if you write the remainder after dividing by 7, instead of 10?

You need some spotty paper and a ruler.

The area of a shape made on pinboard can been seen in two ways.

 **the sum of smaller shapes**

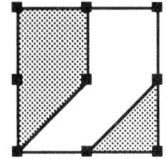 **what is left when pieces are cut off a larger shape**

1 On the spotty paper, make some shapes with no pins inside them.

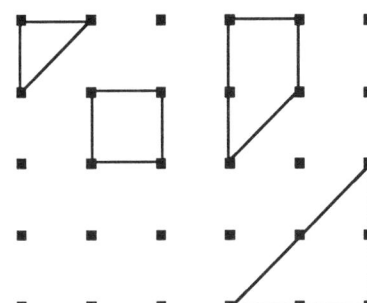

Start with three pins on the edge of your shape, then four, then five, and so on, up to shapes with 25 pins on the edge.

Copy and complete this table.

Number of pins on edge (B)	3	4	5	6	7 . . .
Area in square units (A)	0.5	1			

Find a relationship between the number of pins on the edge and the area of the shape. Write it as a formula.

2 Now make different shapes each with eight pins on the edge. Start with no pins inside, then one pin, two pins, and so on. Copy and complete this table.

Number of pins inside (P)	0	1	2	3	4 . . .
Area in square units (A)	3				

Find the relationship between the number of pins inside the shape and its area. Write is as a formula.

What if there were ten pins on the edge?

1 You need some squared paper and a ruler.

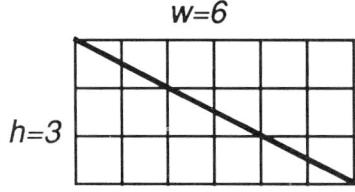

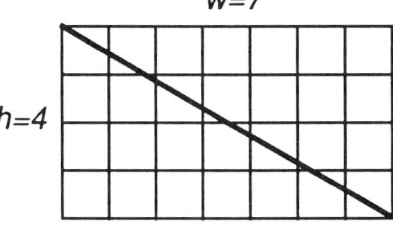

On squared paper, draw lots of rectangles of different sizes.

The width (*w*) and the height (*h*) should each be a whole number.

Draw the diagonal of each rectangle.

Count the number of squares (*s*) which the diagonal crosses.

Make a table.

w	*h*	*s*
6	3	6
7	4	10

There is a relationship between *w*, *h* and *s*. See if you can find it.

Write it as a formula.

2 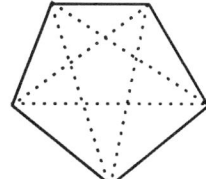 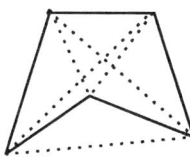 A diagonal of a polygon is a straight line joining any two vertices (corners) not next to each other.

a. How many diagonals are there in:

     i. a quadrilateral   ii. a pentagon   iii. a hexagon?

b. How many in a polygon with *n* sides? Write it as a formula.

c. How many diagonals can be drawn, without crossing each other, in:

     i. a pentagon   ii. a hexagon   iii. a heptagon.

d. How many in a polygon with *n* sides? Write it as a formula.

1 Some bees made a honeycomb.
They started on day 1 with the middle cell.
Each day they added another ring of cells
all round the honeycomb.

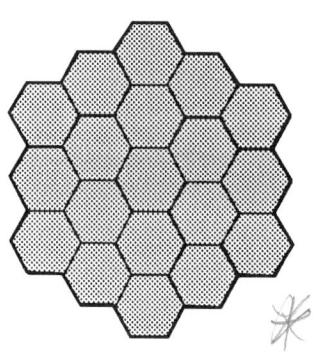

How many cells were there after the sixth day?
On which day were there more than 1000 cells?

2 You need some squared paper.
Some 2 x 1 rectangular tiles are used to make paths 2 units wide.

How many different ways can you make a path four units long?
Five units long?   Six units long?   Longer?

3 You need a tape measure marked in inches and a long strip of
paper.

With the help of the tape measure, make a paper strip one yard long.
Mark distances of 1, 3, 6, 13, 20, 27, 31 and 35 inches from one end.

Investigate the lengths which can be measured by using the strip.

What is the least number of marks needed on a one foot strip so it
can measure every whole number of inches from 1 to 12?

Co-ordinates 1

You need some squared paper, a red pen and a blue pen.

1 Play **Odds** with a friend.

Mark out and label 4 x 4 grids.

Block out one square so it cannot be used.

Take turns to go first.

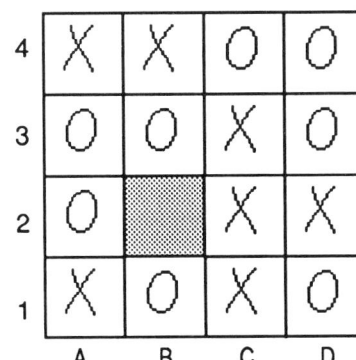

Rules

• Decide who is O and who is X.

• Take turns to tell the other player where to

   put up to three marks.

• Only one mark can be put in each square.

• When the grid is full, each player counts up his or her marks.

The winner is the player who has an **even** number of marks.

2 Play **Squares** with a friend.

Mark out some 6 x 6 grids.

Number the axes.

Rules

• One player is red, the other blue.

• Take turns to tell the other player

   where to put a dot: eg, at (2, 4).

• The winner is the first player to see

   four dots of their own colour at the four corners of a square.

Play several times. Take turns to go first.

3 Play **Squares with a difference**.

Continue playing until the board is full.

Each player records the co-ordinates of each of their squares.

The winner is the player with the larger number of squares.

1 You need a large piece of squared paper.

In the game of **Target** you can make any of three moves:

      either   three spaces East and two spaces North;

      or   five spaces East;

      or   four spaces East and five spaces North.

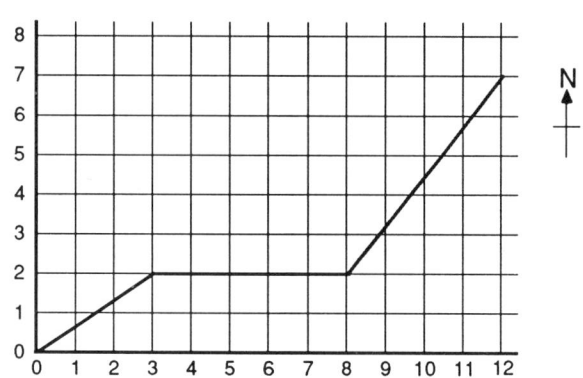

Start at (0,0). How close can you get to (30,40)?

Write the co-ordinates of the points on your path.

Choose another target. Investigate ways of reaching it.

2 You need some squared paper and one pen.

Investigate **Force three** with two friends.

Mark out some grids of 7 x 7 squares.

Number each axis from 0 to 7.

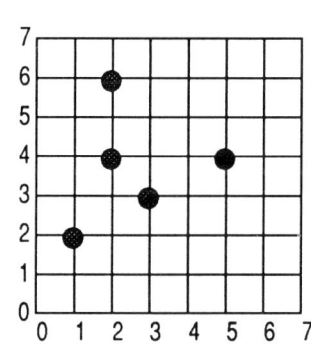

Rules

• One person has the pen and is the marker. The others are players.

• In turn, players name a point on the grid: for example, (3,2).

• The marker puts a dot on that point.

• The **loser** names a point making three dots in any straight line, either horizontal, vertical or sloping.

What is the maximum number of counters which can be played?

1 You need some squared paper and two different coloured pens.

Play the game of **Make five** with two friends.

Mark out some grids of 7 x 7 squares.

Number each axis from 0 to 7.

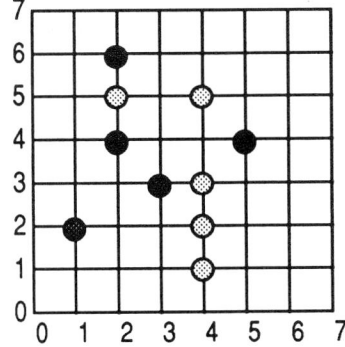

Rules

- One person, the marker, has the two pens.
- Each of the two players chooses a colour.
- In turn, players name a point on the board.
- The marker puts a dot on that point in the player's colour.

The winner is the first to make a straight line of **five** of his or her dots, in any direction.

Write in order the co-ordinates of the five points in each winning line. What do you notice?

2 You need some centimetre squared paper and a ruler marked in mm.

On squared paper draw and measure some squares. Complete this table.

Length of side (cm)	1	2	3	4	5
Length of diagonal (cm)					

Make a line graph to show your results.

Remember to label the axes.

Use the graph to estimate the diagonal length of these squares:

    a.   2.5cm x 2.5cm   b.   4.5cm x 4.5cm

1 You might need some squared paper and a ruler.

The Logo command SETPOS [250 -100] moves the turtle to the position (250,-100).

If PENDOWN, the turtle will draw a straight line as it moves.

This Logo procedure will draw a triangle.

```
TO TRIANGLE1
    PENUP
    SETPOS [300 -100]
    PENDOWN
    SETPOS [-50 250]
    SETPOS [-200 -250]
    SETPOS [300 -100]
    PENUP
END
```

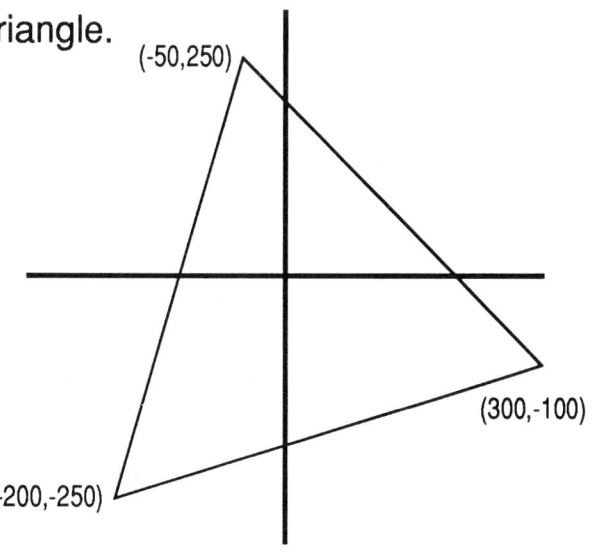

Draw a smaller triangle inside the larger one.

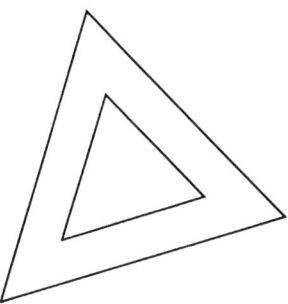

Draw a star made from two triangles.

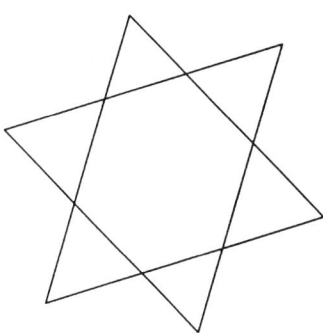

Draw a square to touch each vertex (corner) of the triangle.

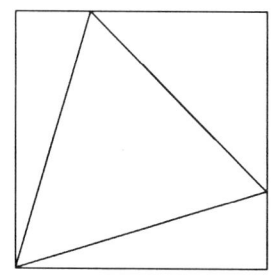

Make up your own problems using the SETPOS command.

1 You need some squared paper and a ruler.

The pair of numbers p = 2 and q = 7 fit this equation.

$$p + (2 \times q) = 16$$

Find other pairs of numbers
which fit the equation.
Record them in a table.

p	2				
q	7				

Plot the set of number pairs on a graph.
What do you notice?
Use the graph to help you find three
more number pairs which fit the
equation p + 2q = 16.

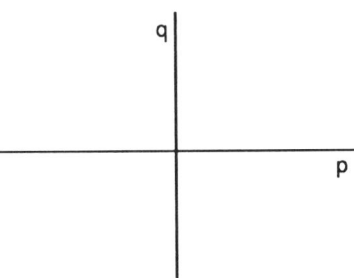

Now find some pairs of numbers to fit this equation.

$$(3 \times p) + q = 13$$

Record the number pairs
in a table.

p					
q					

Plot the second set of number pairs on the same graph.

2 Use your graph to solve this problem.

At the sweet shop, three chews and one liquorice stick cost 13p.
One chew and two liquorice sticks cost 16p.

What is the cost of:

    a.   a stick of liquorice;

    b.   one chew?

1 You need some cubes that fit together.

Make two shapes like this.   They fit together to make a cube.

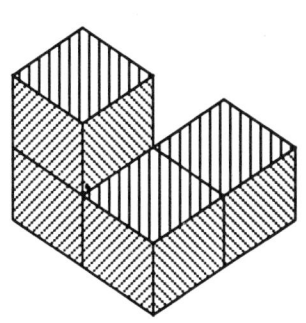 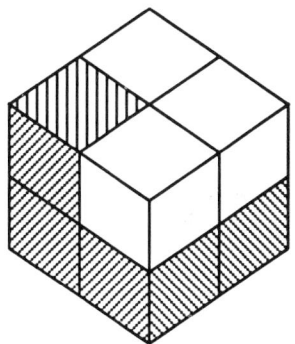

Investigate other ways of making a 2 x 2 x 2 cube from two shapes which are both the same.
How many more ways did you find?

Investigate ways of making a 2 x 2 x 2 cube from two shapes which are both different.

2 A cube has edges 4cm long.
How many 1cm cubes were needed to make it?

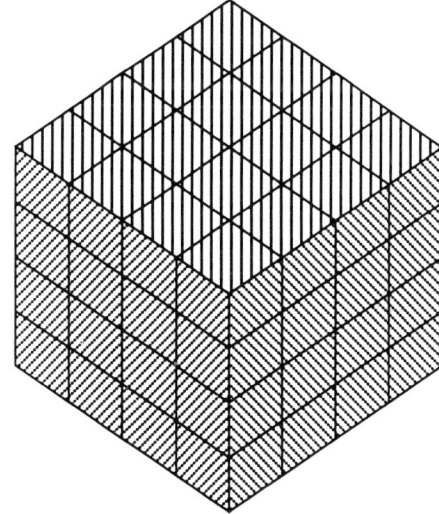

How many 1cm cubes are on the outside of the cube?

Investigate for larger cubes.

1   23 cubes have been fitted
together face to face.

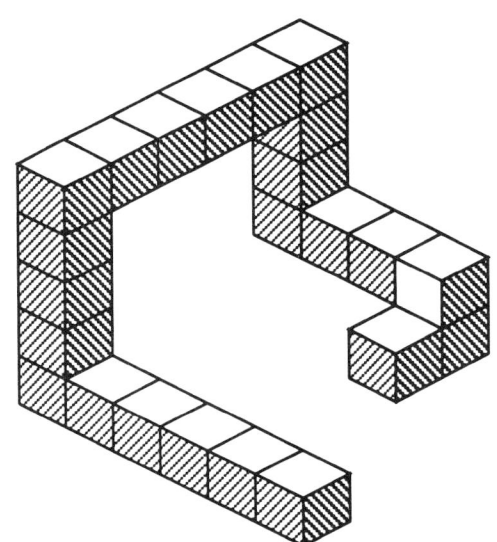

What is the least number of
cubes needed to complete
the loop?

2   You need some squared paper.

Imagine a cube with one black face which is on the table.

There is a flag shape on
each of the four sides like
this: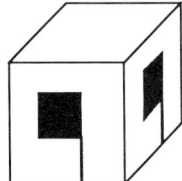

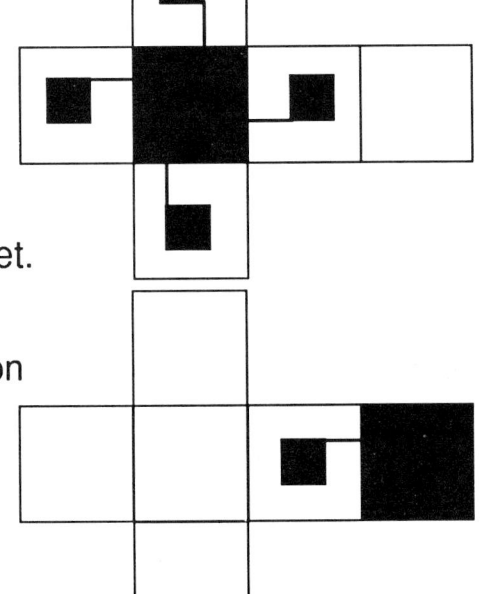

The cube is made by folding up this net.

Where would you put the other flags on
this net so it makes the same cube?

3   You need some squared paper and scissors.
Hexominoes are made from six squares which touch edge to edge.
Ten different **hexominoes** will fold to form a cube.
Which are they?

3D shape 3

1 These shapes can be folded to make three dice.

Number the squares so that opposite faces add up to 7.

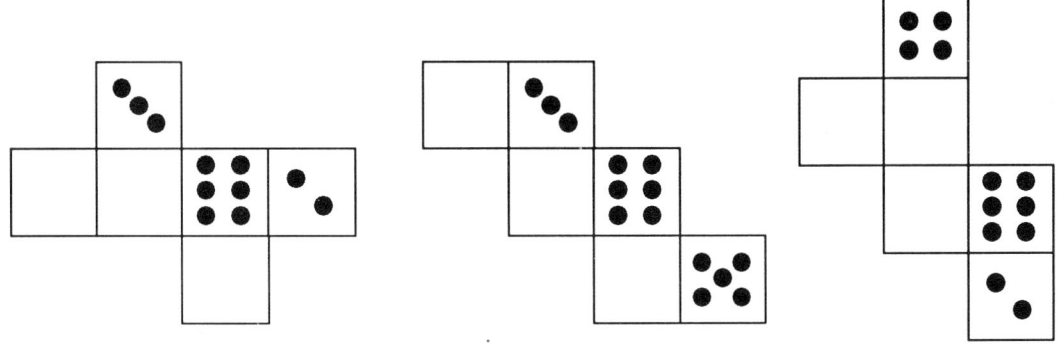

2

This parcel is tied with three bands of ribbon.

Draw the bands on the three nets so that they will make the same parcel when folded.

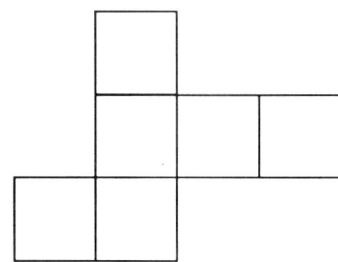 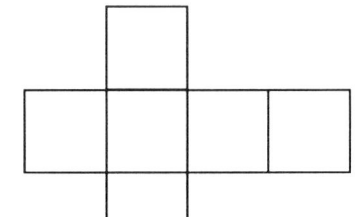 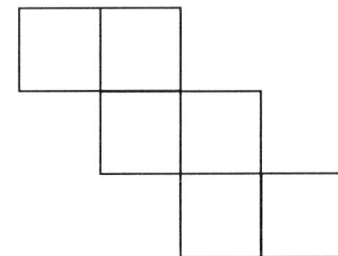

3 Imagine a 3 x 3 x 3 cube.
The outside is painted red.

How many of the unit cubes have:

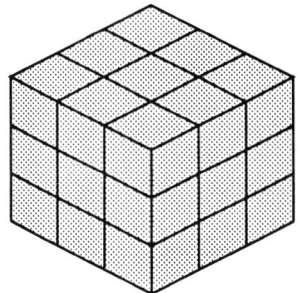

0 red faces;

1 red face;

2 red faces;

3 red faces;

4 red faces?

What if you started with a 4 x 4 x 4 cube? Or a 5 x 5 x 5 cube?

1 You need: squared paper;

9 red cubes, 9 green cubes and 9 blue cubes.

You can use different colours but you will need nine of each.

Build a large cube. Each row and each column in each face must have all three colours in it.

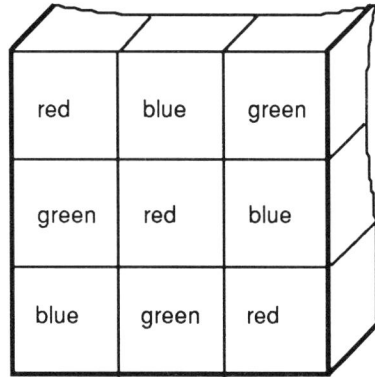

Make some other large cubes.

Draw the pattern of the colours on each of the faces.

2 You need four cubes.

Here are two arrangements of three cubes.

In each shape the cubes touch each other face to face.

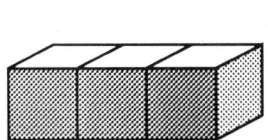 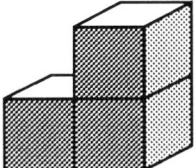

Count the number of squares on the outside of each shape.

Do both shapes have the same number?

Investigate shapes made from four cubes.

How many different shapes can you make?

Do they all have the same number of squares on the outside?

Can you explain why?

You need some plain paper, an equilateral triangle, a ruler, some scissors, and some coloured pens (one red, one yellow, one blue, one green).

1 Draw round the equilateral triangle.
Cut it out.
Draw lines on the triangle to divide it into four smaller triangles.
Colour the big triangle using a different colour for each small triangle.

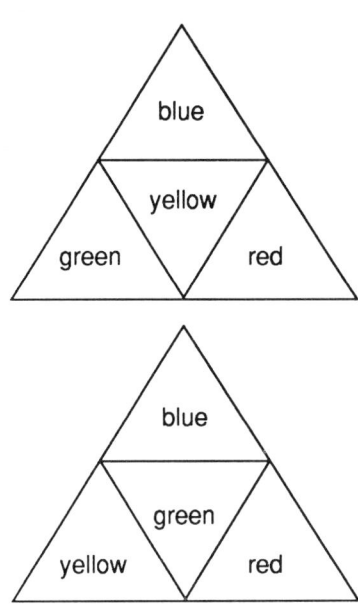

How many different ways of colouring the big triangle can you find?

What if you only used two colours?
You can use one of them or both of them.
How many different ways of colouring the big triangle are there?

Fold each four-colour triangle to make a tetrahedron.
Each face of the tetrahedron is a different colour.
How many different tetrahedra are there?

What if you fold the two-colour triangles?
How many different tetrahedra are there?

2 How many edges of a cube must be cut so it can be folded flat without falling to pieces?

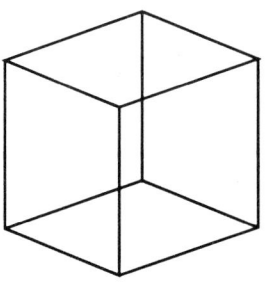

Is it always the same number?

1 You need some Multilink cubes.

Make some shapes from five cubes touching face to face.

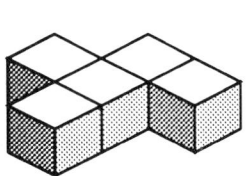 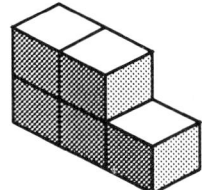

These shapes lie in just one plane.   **These shapes lie in two planes.**

How many different shapes lying in two planes can you make?

Which has the least surface area?

2 Imagine you have eight wooden cubes,
some red paint and some blue paint.

How could the eight cubes be painted so that
they would fit together to make:
either a red 2 x 2 x 2 cube,
or a blue 2 x 2 x 2 cube?

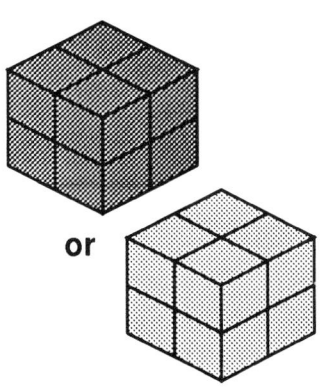

or

3 You need some squared paper, some scissors, and coloured pens.

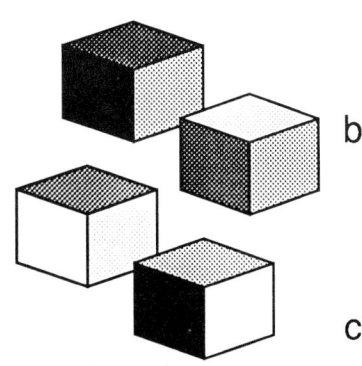

a. A cube is painted so that faces which are
next to each other are different colours.
What is the least number of colours needed?

b. If four colours are used, and faces next to
each other are different, how many different
cubes can be made?

c. How many differently coloured cubes can
you make using six colours, one colour for
each face?

1 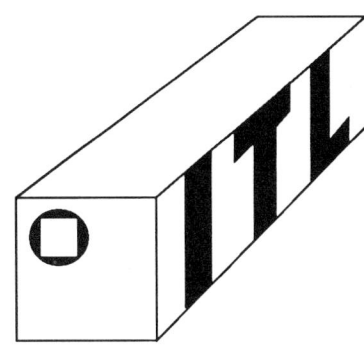 Some containers for transporting goods by road come in three lengths: 12m, 9m and 6m. Each size is 2.3m wide and 2.3m tall.

Crates are to be loaded into the containers. Each crate measures 1.1m x 1.1m x 2.9m.

    a.   How many will fit in a 6m container?

  b.   How many will fit in one 12m and two 9m containers?

2 The length of a rectangle is twice its width.

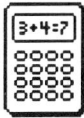

  a.   If its area is 50 square centimetres, what are its measurements?

  b.   What if the area is 20 square centimetres?

  c.   What if the area is 40 square centimetres?

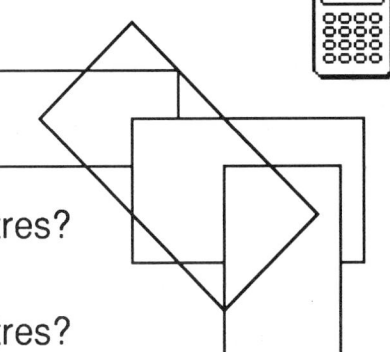

3 Imagine making an open box from a square of cardboard. You could cut out square corners and fold up the sides.

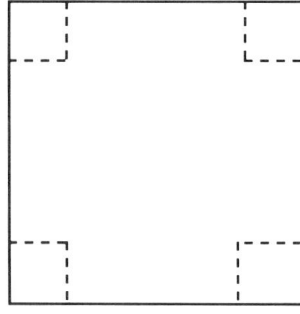 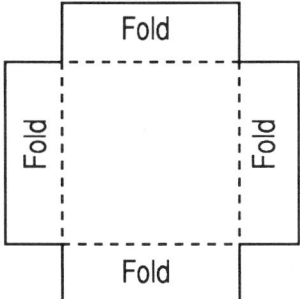

  a.   If you started with a 40cm x 40cm piece of card, what size corners should you cut so the box has the biggest volume?

  b.   What if you started with a 30cm by 20cm rectangle?

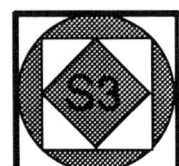

1 How many triangles can you find?

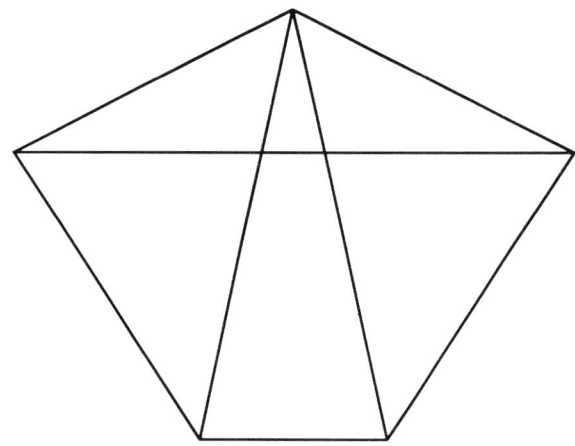

2 How many rectangles in each of these?

a.	b.	c.	d.	e.

Can you see a pattern?

How many rectangles in the next diagram?   And the next?

3 You need 20 straws all the same length.

Arrange 12 straws to make five squares. You must not bend them!

Draw the pattern.

How many squares can you make using 20 straws?

4 Only one of these statements is true. Which one is it?

All squares are rectangles.

All rectangles are squares.

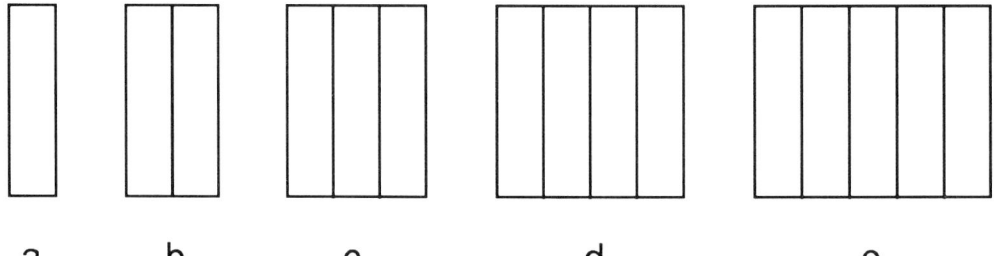

1 You need some squared paper.

Investigate shapes made from squares that touch edge to edge.

How many different shapes can you make from 4 squares?

How many different shapes can you make from 5 squares?

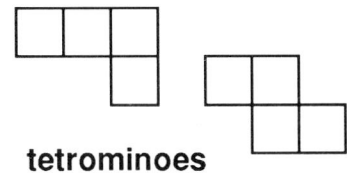

tetrominoes

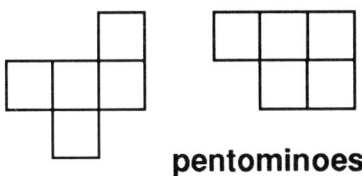

pentominoes

What other words start with **pent** or **tetr**?

2 You need some squared paper.

Which of your tetrominoes will tessellate?

Draw and colour some patterns to show how.

Which tetrominoes will tessellate to form a 4 x 4 square?

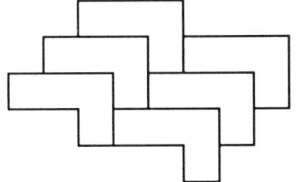

3 One of the pentomino shapes will tessellate.

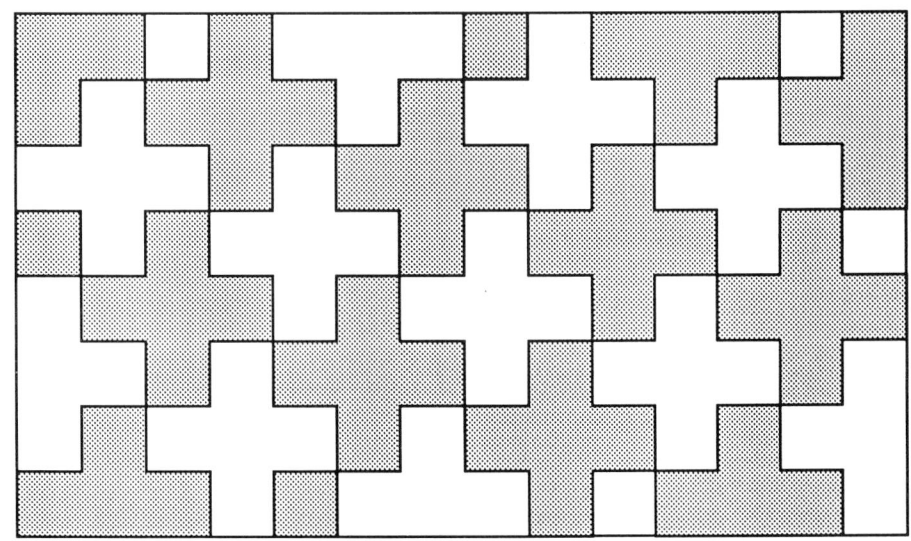

Which of the other pentomino shapes will tessellate?

Draw and colour patterns to show how to do it.

1 You need squared paper and scissors.

Pentominoes are made from five squares touching edge to edge.

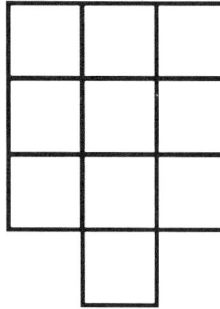 Divide this shape into two different pentominoes.
In how many different ways can you do this?
Draw them on squared paper.

Fit two pentominoes together. Draw round them to make a new shape. Divide the new shape into two different pentominoes. Investigate different ways of doing it.

Find a shape into which only two pentominoes will fit.

2 You need 6 straws all the same length, some scissors and a ruler.
Measure one of the straws in centimetres.
Cut two of the straws in half.

You now have 8 straws, four long and four short.
Use the 8 straws to make three squares,
all the same size.
You must not bend the straws!
Draw the pattern.
What is the area covered by all three squares?

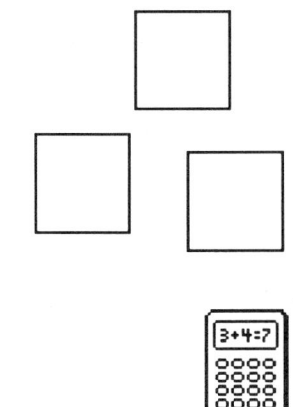

3 You need some squared paper.

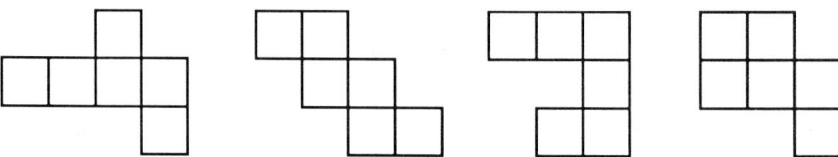

Draw as many as possible of the 35 different **hexominoes**.

You need some spotty paper.

1 The area of this square is 5 square units.

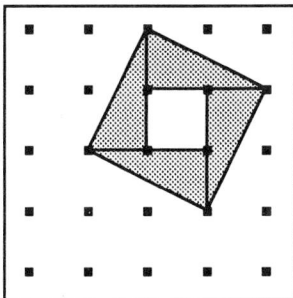

It is made from four triangles, which are half rectangles, plus one small square in the middle.

On 5 x 5 pinboard, how many squares with an area of 5 square units can you make?

Investigate squares of different sizes on a 5 x 5 board.
How many different sizes of squares can you make?

2 Using 5 x 5 pinboard, make triangles with area $\frac{1}{2}$, 1, 1$\frac{1}{2}$, 2, 2$\frac{1}{2}$, ... square units, and so on.

What is the area of the largest triangle you can make?

How many different triangles with the largest area can you make?

How many different shaped triangles with area $\frac{1}{2}$ square unit can you make?

3 Each of these shapes is two square units in area. Draw some more.

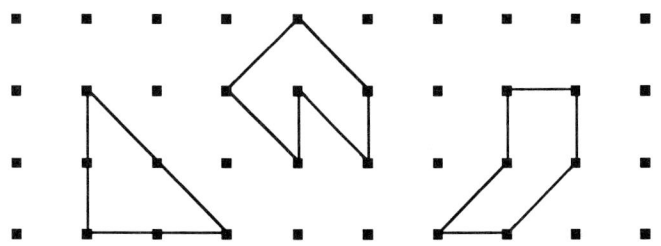

Investigate ways of covering completely a 5 x 5 pinboard with shapes, each of area 2 square units. Each shape must be different.

1 The first square has been cut into 18 square pieces.

Investigate ways of cutting the other squares into square pieces.

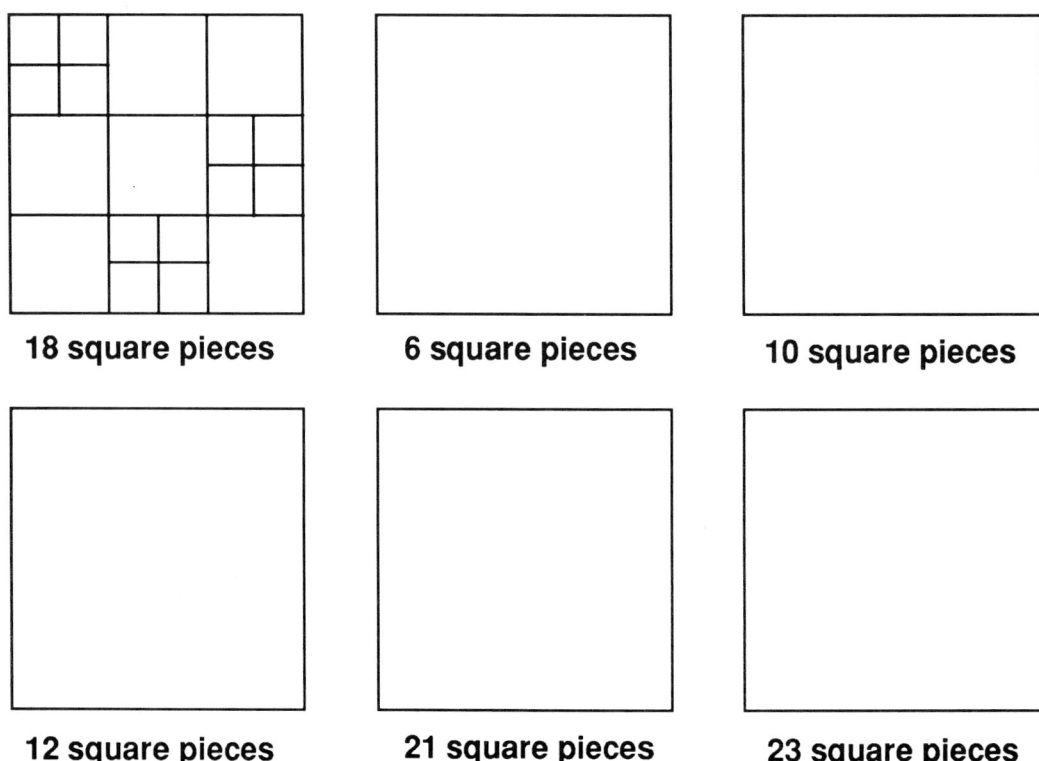

18 square pieces   **6 square pieces**   **10 square pieces**

12 square pieces   **21 square pieces**   **23 square pieces**

Which numbers of square pieces are impossible?

2 You need some tracing paper, a ruler and some scissors.

Make a tracing of this fish. With your ruler, mark just two straight cuts.

Cut out the pieces and reassemble them to make a square.

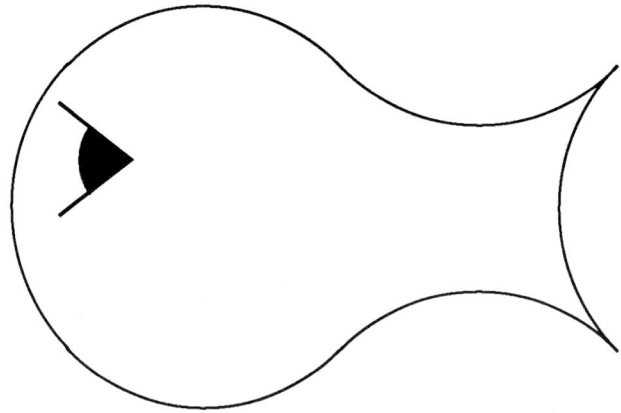

1 This spiral is made by drawing arcs of quadrants inside squares.

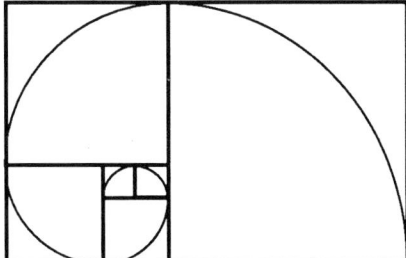

The smallest square is 1cm x 1cm. What is the length of the spiral?

2 These patterns are made with circles, semi-circles and quadrants inside squares of the same size.

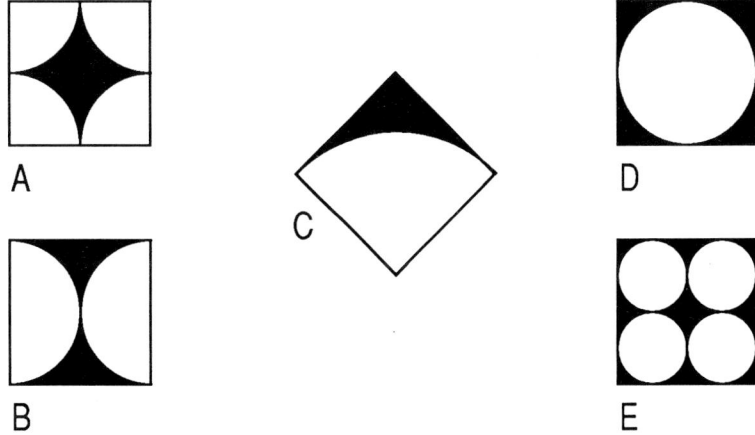

a. Which pattern has the greatest black area?

b. Boundaries between black and white areas are curved lines. Which pattern has the greatest length of curved line?

3 Each of these wheels is fixed at its centre and rotates without slipping. Wheel A is 4cm in diameter. It rotates 12 times each minute.

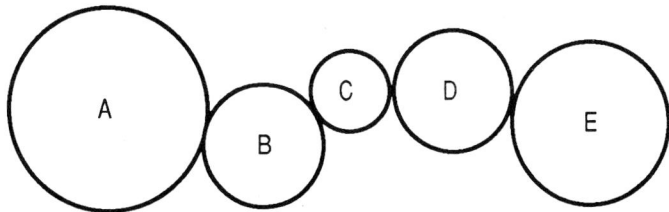

Wheel E is 3cm in diameter. How fast does it rotate?

1 A rectangular box has a top area of 120 square centimetres.

The area of a side is 96 square centimetres.

The area of an end is 80 square centimetres.

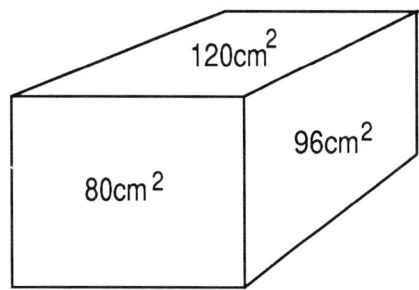

What are the exact measurements of the box?

2 Investigate the smallest rectangular area of wrapping paper which can be used to wrap this parcel.

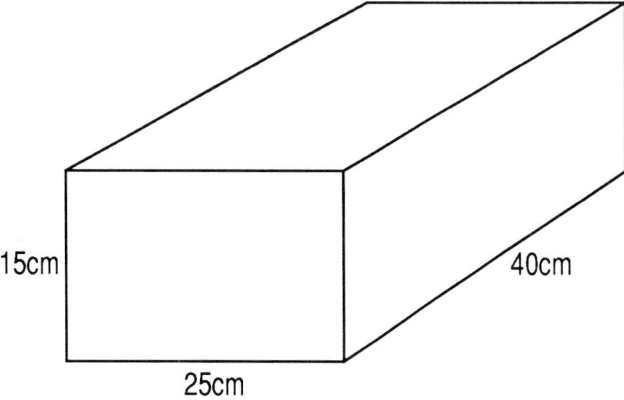

3 You need some sheets of paper, a ruler and some scissors.

Design and make an envelope. Investigate different ways of doing it.

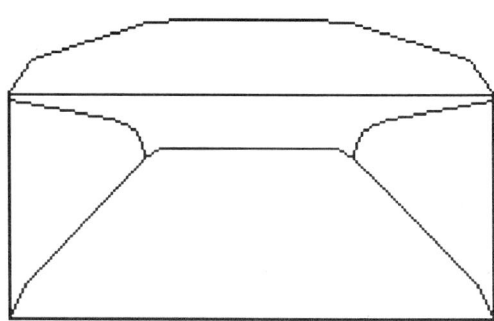

1 You need some spotty paper.

parallelogram 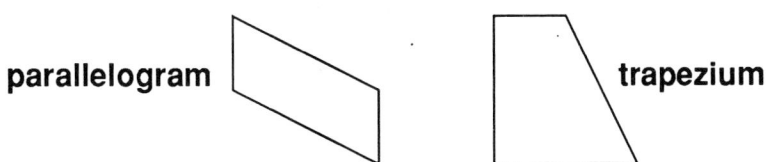 trapezium

A **parallelogram** has its opposite sides equal and parallel.
Using 3 x 3 pins, how many different parallelograms can you make?

A **trapezium** has one pair of parallel sides.
Using 3 x 3 pins, how many different trapeziums can you make?

Investigate ways of making different polygons with 5, 6 or 7 sides.
Which of your polygons will tessellate?

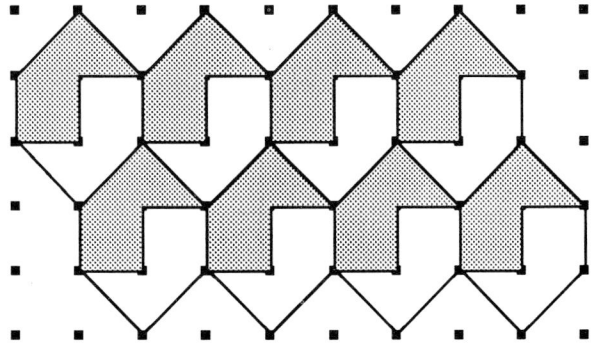

2 You need some short straws, all the same length, and some longer straws, all the same length.

How many different triangles can you make?

What about quadrilaterals?

What if you had straws of three different lengths?

1 You need some squared paper and four coloured pens.
Draw this pattern. Make each circle a different colour.

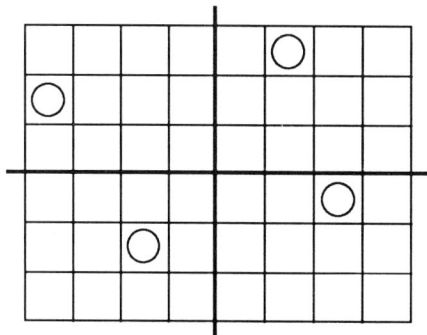

The pattern has two lines of symmetry. Finish the pattern.

2 You need plain and squared paper, a ruler, scissors and glue.

Draw some 3 x 3 tiles.
Shade three small squares to make a pattern.

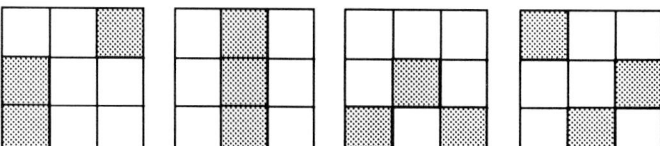

How many different 3 x 3 tiles can you make?

Use your ruler to draw any lines of symmetry.
There could be more than one line.

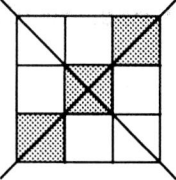

Cut out the tiles that are not symmetrical.
Can you put any of them together
to make a symmetrical pattern?
Stick them side by side.

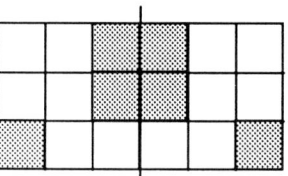

Make some repeating patterns using the 3 x 3 tiles.

1 How many triangles are there in each of these figures?

 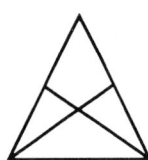 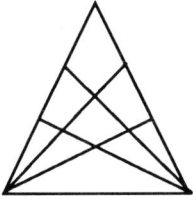

Draw the next figure in the sequence.

How many triangles are there in it? How many in the next figure?

2 You need a ruler, a set square and a sharp pencil.

Here is part of a lattice design. It is a traditional form of Chinese art.

The lattice has four lines of symmetry. Can you complete it?

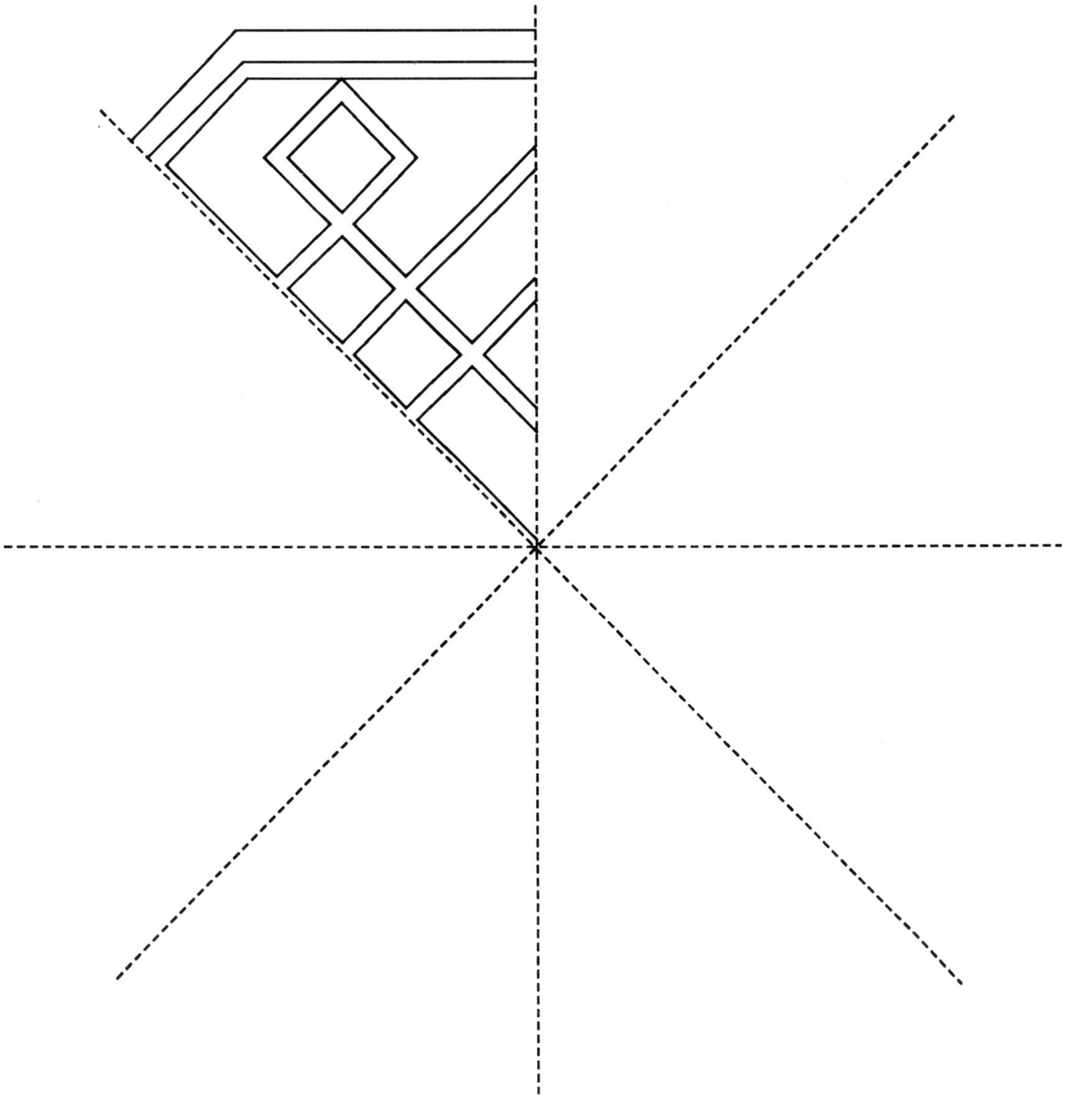

1 The straight lines represent mirrors.

Draw the reflections of each letter.

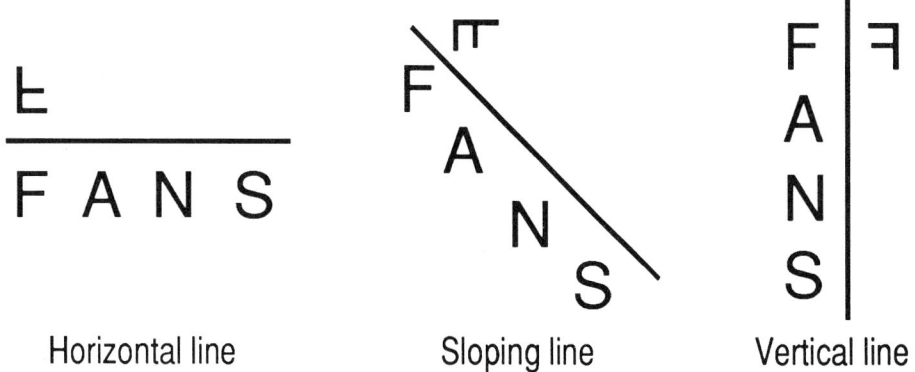

Horizontal line   Sloping line   Vertical line

Investigate for other capital letters.

Which letters have a mirror image which looks the same as the original letter?

2 Draw the letter F if it were turned round upside down like this:

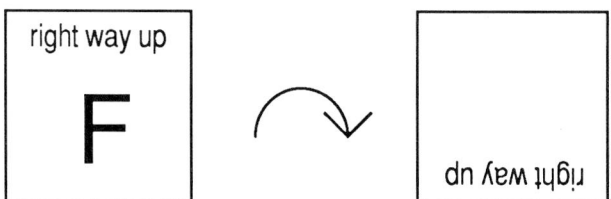

Which capital letters when rotated through half a turn have an image which looks the same as the original?

3 You need a potato and a tool for cutting it.

Make a potato stamp which will print a letter F.

The library uses a rubber stamp to print dates on books like this.

JAN 13

One day the librarian noticed that the date on the stamp itself seemed the right way round, not back to front or upside down. What was the date?

1 The sides of the big triangle are 5, 6 and 7 units long.
The middle points of each side are joined
to make a new triangle.

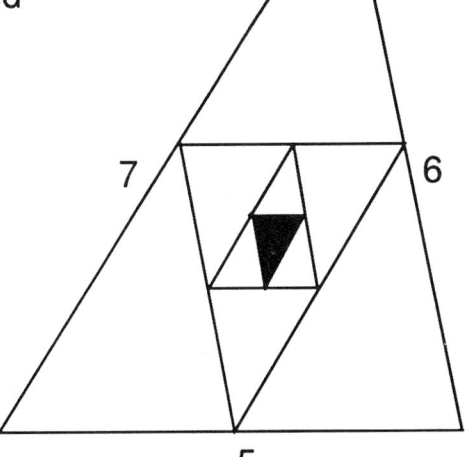

a. How long are the sides of the
black triangle?

b. What fraction of the big triangle
is the area of the black triangle?

2 You might need a ruler.

The diagram shows two maps of the same area of land.
The larger map is twice the scale of the smaller map.
There is a point on the smaller map which is exactly above the
same point on the larger one.
Can you find it?

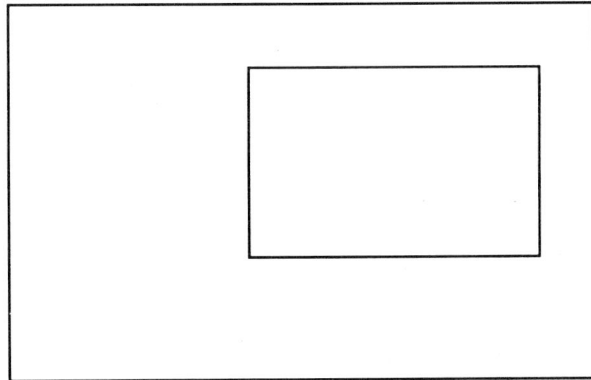

3 These Logo commands will make the turtle draw a circle.

```
REPEAT 36 [FORWARD 15 RIGHT 10]
```

Investigate ways of using Logo to draw bigger or smaller circles.

1

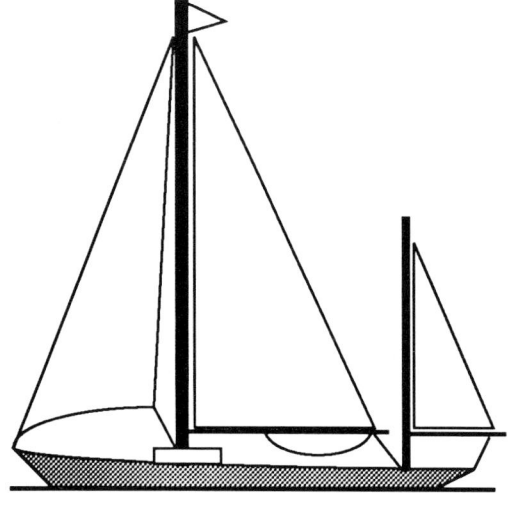

You need some suitable paper, a ruler and a set square.

Draw this sailing boat three times the size it is here. How could you make your enlargement more accurate?

Make up more problems like this.

2 Jo measured the lengths of a vertical stick and its shadow. At the same time she measured the shadow of a tall building.

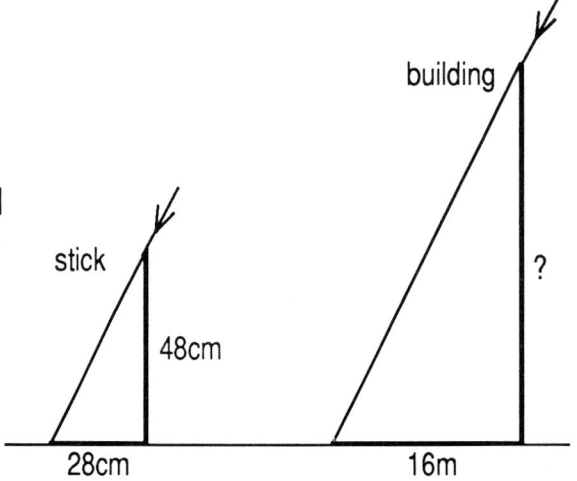
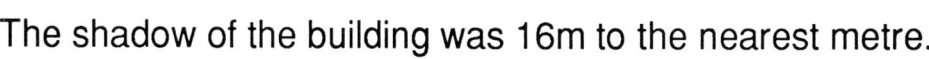

stick

building

48cm

?

28cm

16m

The stick was 48cm to the nearest centimetre.
Its shadow was 28cm to the nearest cm.
The shadow of the building was 16m to the nearest metre.

Using what you know about the accuracy of these measurements, find the greatest and least possible height of the building.

What would Jo have to do to find the height more exactly?

3 Find the approximate heights of some parts of your school building.

1 You need some spotty paper.

Put a small cross in the centre.

You can only move north, south, east or west.

You can only hop one dot at a time.

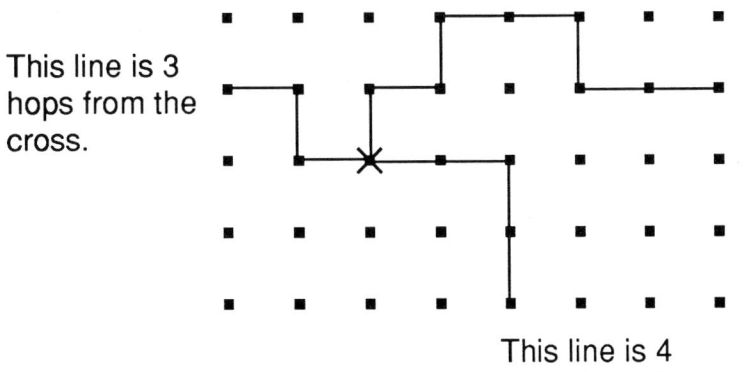

This line is 3 hops from the cross.

This line is 8 hops from the cross.

This line is 4 hops from the cross.

Find all the dots that are four hops from your cross.

What shape do they make?

Investigate for other hopping distances.

2 Start here at the lake. Move one dot at a time.

Draw this path: N, E, S, S, W, W, N, N, N, W.

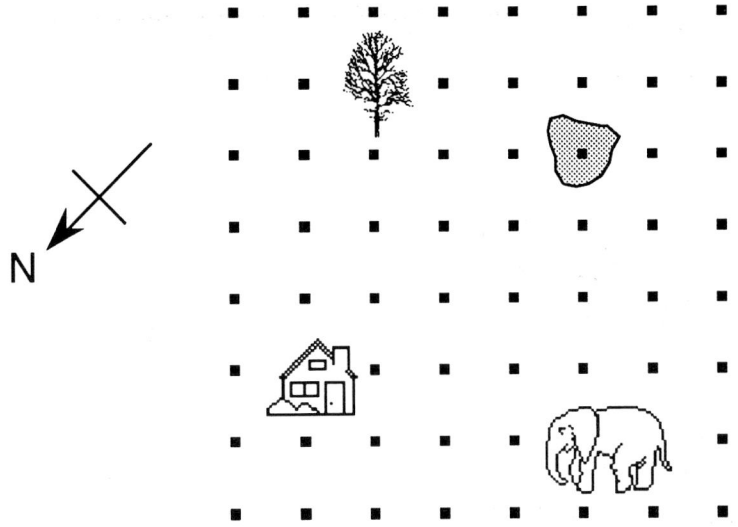

Describe a path from the lake, round the house, to the tree.

Ask your friends to try it.

1 You need some spotty paper.

Investigate the triangles which you can make on 3 x 3 pinboard.

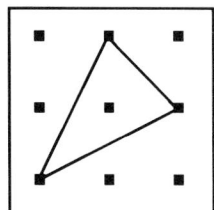 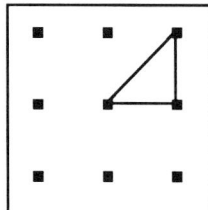 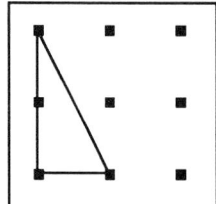 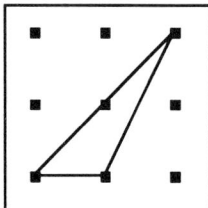

How many different triangles can you make?

How many of these have a right angle?

2 You need some squared paper.

Start near the middle.

Do this procedure.

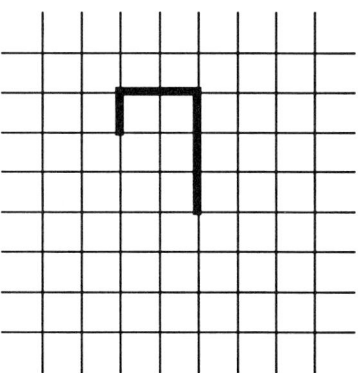

- Draw a line 1 unit long.

- Turn a right angle clockwise.

- Draw a line 2 units long.

- Turn a right angle clockwise.

- Draw a line 3 units long.

- Turn a right angle clockwise.

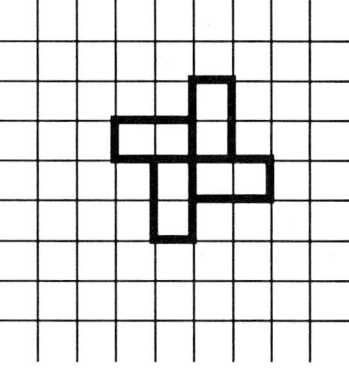

Keep repeating the procedure.

How many times did you repeat it?

What happened?

That was a (1, 2, 3) pattern.

Here is a (3, 1, 1) pattern.

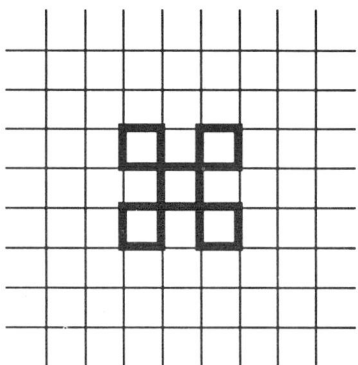

Investigate for other numbers.

Try using Logo.

1

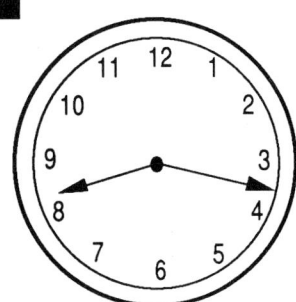

At 8:18, the angle between the hands of the clock contains 4, 5, 6, 7 and 8, which add to 30.

The reflex angle between the hands contains 9, 10, 11, 12, 1, 2 and 3 whose sum is 48.

What time do the angles between the hands contain sets of numbers with equal totals?

2 The dial of a machine for printing numbers on tickets has the ten digits 0-9 evenly spaced around it.
You turn the dial in either direction until the digit you need is opposite the arrow.
The ticket moves along automatically.
The machine starts each ticket with the digit 0 opposite the arrow.

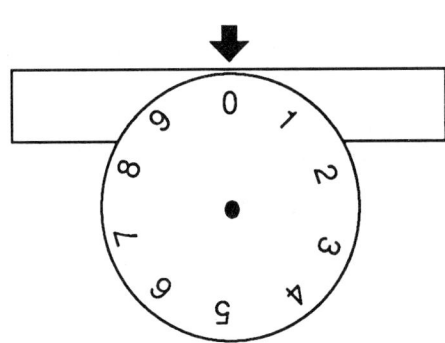

What is the smallest total angle which the dial must be turned to print these numbers?

     a.   2084   b.   51930

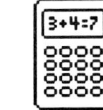

3

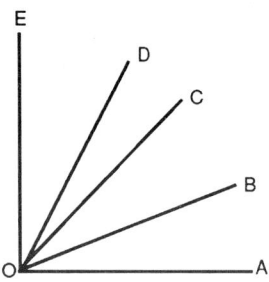

How many angles are there in this diagram? Don't forget reflex angles.

Investigate for other numbers of lines.
Can you see a pattern?

How many angles would be made by 20 lines?

Using five lines, what angles between them would make the least number of different sized angles?

1 You need some spotty paper.

Draw this quadrilateral.

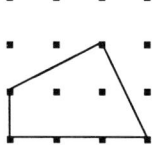

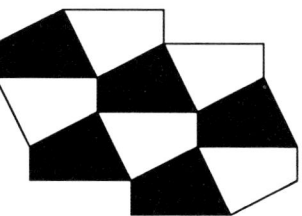

The quadrilateral makes interesting tessellations.

Each of these shapes is made from four of the quadrilaterals.

Make some tessellation patterns with them.

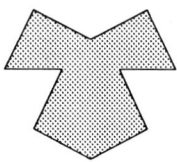 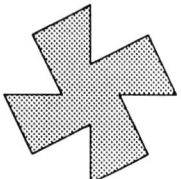 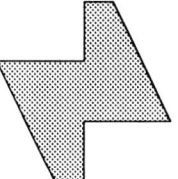

Design some tessellations of your own using the quadrilateral.

2 You need some plain paper, a ruler and a protractor.

Will this pentagon tessellate? Sketch how it might be done.

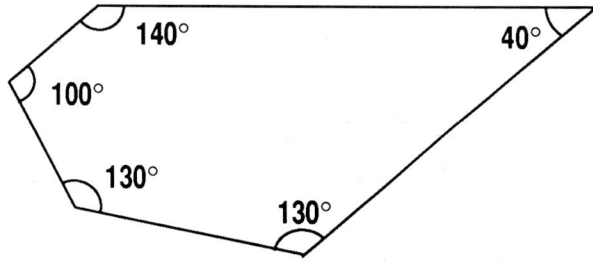

3 This is a tessellation of identical 7-sided polygons.

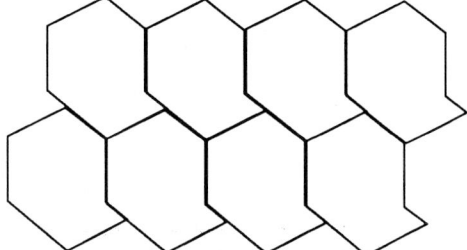

Design some tessellations of 9-sided polygons.

1 You need a ruler and a protractor. On the map, 1cm represents 2km.

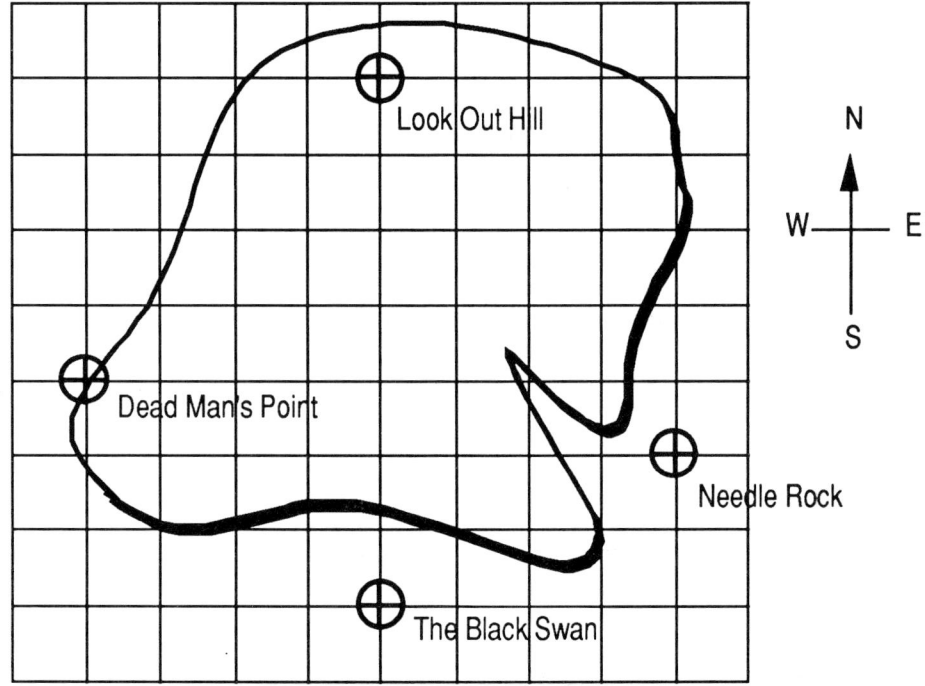

a. What is the actual distance between these points:

    i.   The Black Swan and Look Out Hill;

    ii. Look Out Hill and Dead Man's Point;

    iii. Dead Man's Point and Needle Rock?

b. A telescope on Look Out Hill is pointing due south.
Through what angle is it turned to point at Needle Rock?

c. The guns on The Black Swan are pointing at Dead Man's Point.
Through what angle do they turn to point at Needle Rock?

d. A pirate walks straight from Look Out Hill to Dead Man's Point.
When he gets there he turns to face Needle Rock.
Through what angle does he turn?

Choose a point for some buried treasure and mark it on the map.
Write the distance and bearing of each of the other four points
from your buried treasure.

1 You might need a ruler and a protractor.

The Logo command SETH sets the turtle's heading.

SETH 75 turns the turtle to a bearing of 75°.

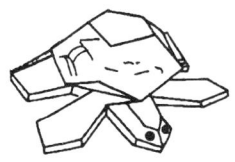

Use Logo to draw a ring of pentagons.

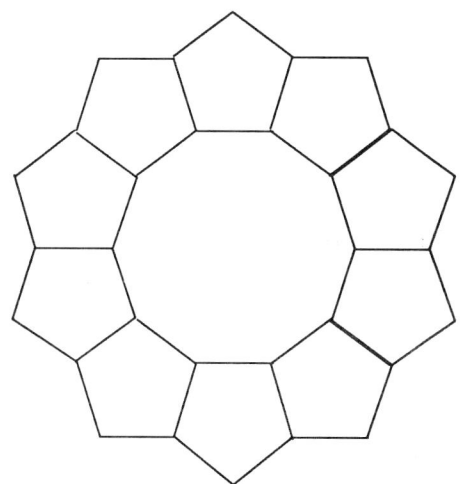

Make up more problems
which you can solve by
setting the turtle's heading.

2 Investigate these Logo procedures.

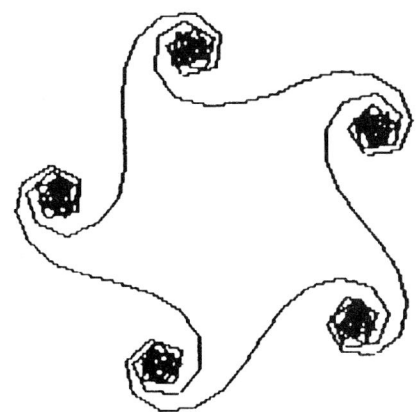

SQUIGGLE 20 33

```
TO SWITCH :length :angle
    FORWARD :length
    RIGHT :angle
    IF HEADING = 0 [STOP]
    SWITCH :angle :length
END
```

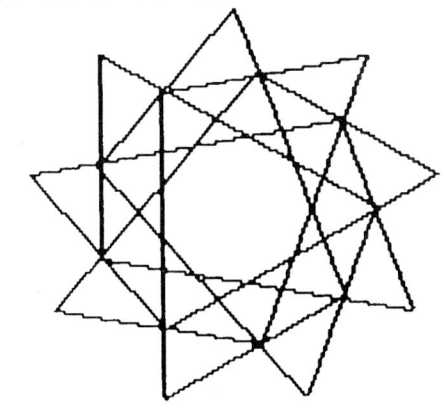

SWITCH 120 80

1

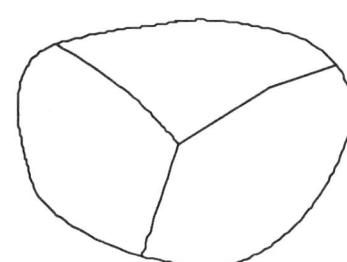

A gardener divided a flower bed into three parts. Each part had a boundary in common with each of the other parts.

Could the flower bed be divided into four parts in this way? What about five parts?

2 A big wheel is made of struts.

A safety inspector has to examine each strut by climbing once along its length.

If he gets on and off the wheel at A, how can he carry out the inspection?

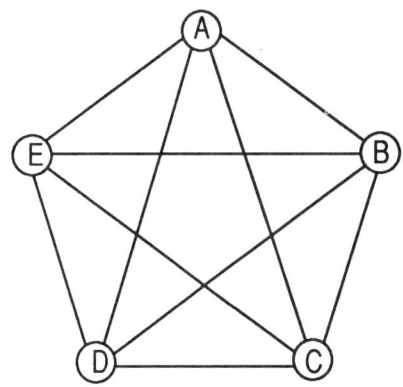

3

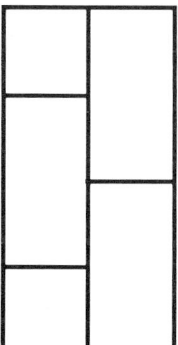

Pam wants to draw this diagram without going over any line twice.
She will have to take her pencil off the paper.
What is the least number of times she will need to do this?

4 Here are six points.
Any two points can be joined by only one line. This line must not cross any other line.
So far, six lines have been drawn.
How many more lines can you draw?

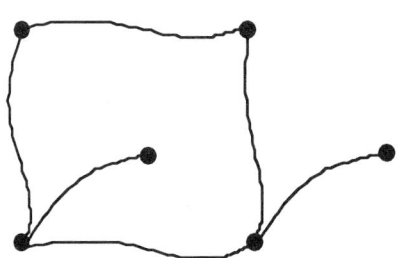

Investigate for other numbers of points.

1 Nine large trees are planted in a park.

The head gardener's office is at point A.

Each week, she checks whether any branches have fallen.

She starts from and returns to her office.

To save time, she prefers not to go along the same path twice,

or to visit a tree more than once.

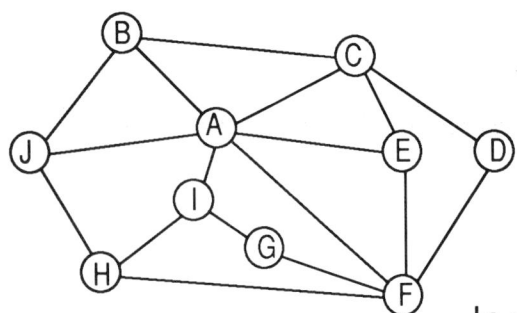

In what order should she check the trees?

2 A police car must patrol each of the 17 roads in this block.

Each road is 1 mile long.

The car can start its patrol at any point and end its patrol at any point. How long is its shortest route?

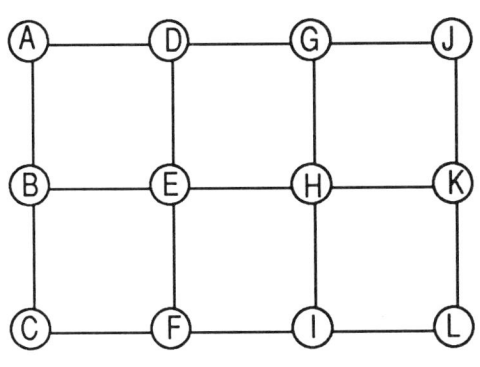

3 In a horse riding competition against the clock, riders must keep to the marked tracks, touching a post at each of the nine points shown, in any order, starting from and returning to A.

The distance in metres between each post is shown. Which route would you recommend:

a. to touch each post at least once;

b. to cover each part of the track at least once?

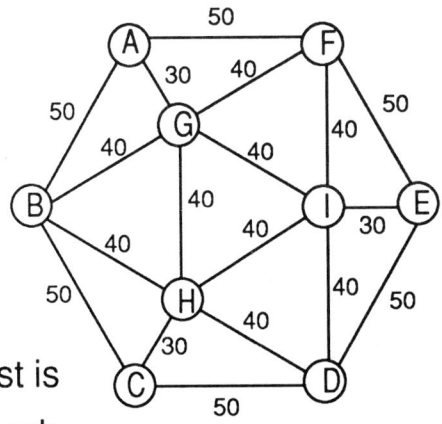

1 Five bridges A, B, C, D and E help travellers cross a river.
Two bridges from each bank cross to an island in the river.

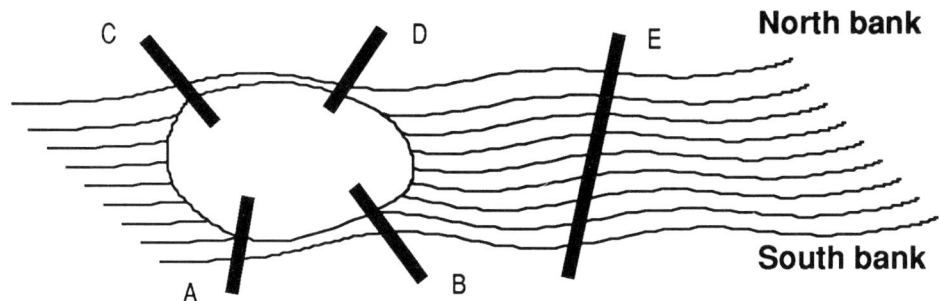

I want to travel from the south bank to the north bank crossing
each bridge once and only once.
How many different routes are there?

2 A mouse can run through all the
rooms in this maze, visiting each
room once and only once.

It enters by the door marked A and
leaves at the door marked B.

How many different routes can it take?

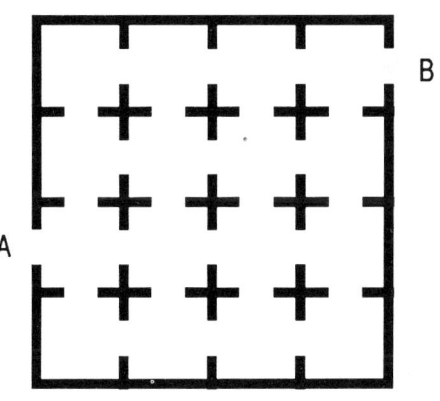

3 The Italian mathematician Guarini invented this puzzle in 1512.

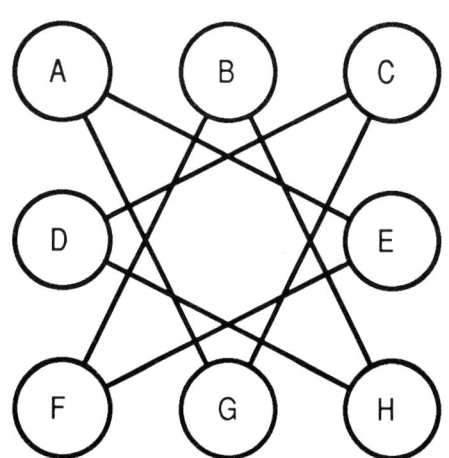

Two green frogs sit on the toadstools
marked A and C.
Two brown frogs sit on F and H.
Each toadstool will hold only one frog.

The frogs jump along the straight lines
until they have changed places.
How many jumps are needed?

1

Birmingham New Street	0941	1006	1106	1236
Birmingham International	0950	1017	1116	1247
Coventry	1012	1031	1129	1301
Leamington Spa	1026	....	1143	1315
Banbury	1046	....	1203	....
Oxford	1107	1121	1224	1354
Reading	1131	1156	1250	1425

What is the fastest train between these stations:

      a.   Birmingham New Street and Reading;

      b.   Coventry and Oxford;

      c.   Oxford and Reading?

2 You need some plain paper and a ruler.

A, B, C, D, E and F are six points for bus stops.

The table shows which points are connected by a direct bus route,
and the time taken between them.

Direct bus route	A to B	B to C	C to D	D to E	E to F	F to A	B to D	B to F	C to E	C to F
Time in minutes	4	3	4	4	3	4	5	3	2	2

Mark the six points on a piece of paper.

Join the points that have bus routes between them.

How many different round trips from A back to A can be made,

if each bus stop is visited once and once only?

Which is the quickest round trip?

The six bus stops are joined by straight roads.

None of the roads crosses another.

Draw a map of the roads.

Find different ways of drawing the map.

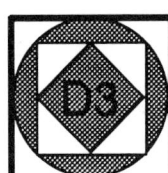

1 You need a small ball or beanbag.

Who catches best?   Boys or girls?

Which hand is best for catching?

Test some girls and boys.

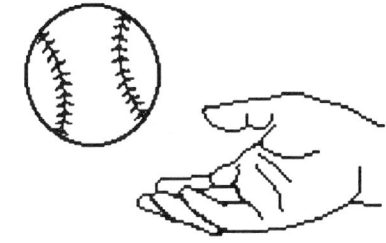

For each one, throw a ball ten times for right-hand catches.

Throw ten more times for the left-hand catches.

Count the catches. Record your results.

Name	Right-hand catches	Left-hand catches

2 You need squared paper and a ruler.

Make a graph of your results.

Show both hands on the graph.

■ right hand

▨ left hand

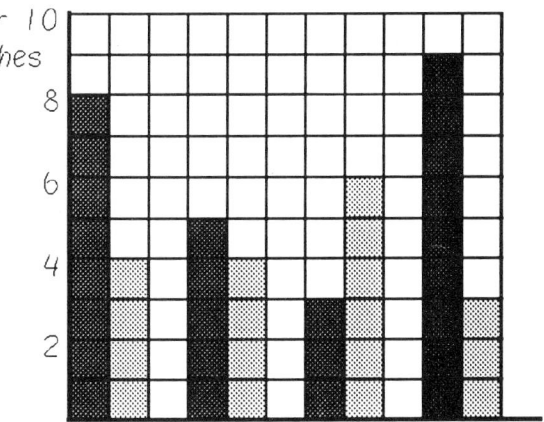

Who catches best?   Which hand is best for catching?   Why?

Would your view be the same if you tested more people?

1 Three rounders teams played against each other.

The table below gives some information about the results.

	Played	Won	Lost	Drawn	Goals for	Goals against
Reds	2		2		2	
Greens	2				3	
Blues	2			1	2	0

Complete the table and find the score in each match.

a.   Reds v Greens   b.   Greens v Blues   c.   Blues v Reds

2 Five football teams formed a league.

Each team played each of the other teams twice.

Points scored were two for a win and one for a draw.

Teams which lost scored no points.

Complete the table below giving information about the results.

Team	Games				Points		Total points	Position
	Played	Won	Drawn	Lost	Won	Drawn		
Rovers	8	4	1	3	8	1	9	
City	8		1	2				
United	8	4			8			2nd
Forest	8	2					6	
Spurs	8	2						

1 You need some plain paper and coloured pens.

The rings enclose four groups of words: **a** , **b** , **c** and **d** .

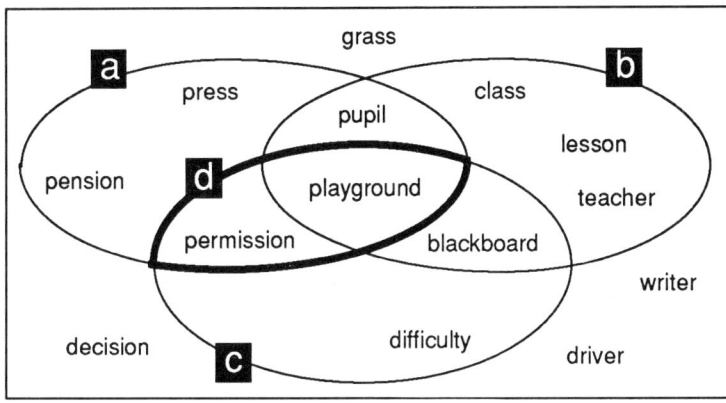

What does each group have in common?

Copy the words again on plain paper, without the lines.

Draw at least three other rings to group words which are alike.

2

Mrs Smith breeds dogs.

She has 11 dogs at present.

7 of her dogs are spaniels. 8 are puppies.

What is the least number of spaniel puppies she must have?

3 Draw the missing figure.

Make up some more problems like this.

Data 4

1 A scientist tested the quality of some peas.

She counted the number of peas in each pod. Here are her results.

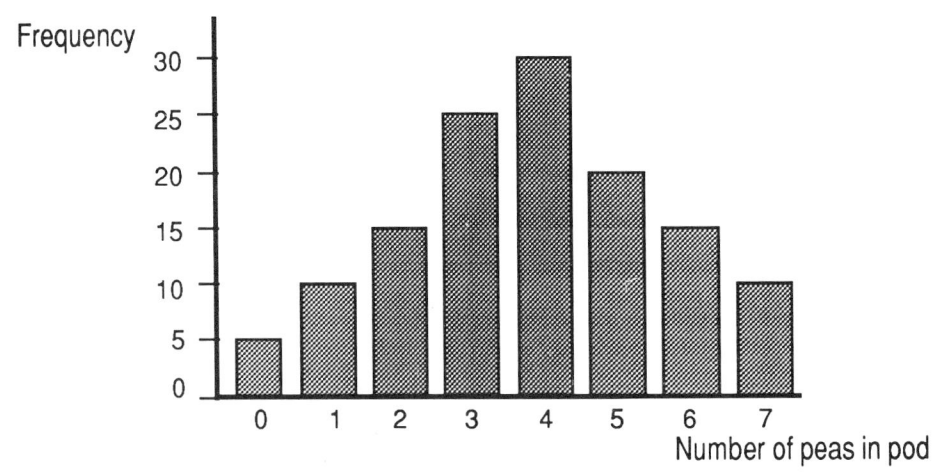

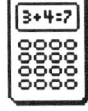

a.   How many pea pods were tested altogether?

b.   What was the most common number of peas in a pod?

c.   How many pods contained just two peas?

d.   What was the total number of peas in pods with two peas?

e.   What was the total number of peas in the test?

f.   What was the average (mean) number of peas in a pod?

2 You need a stop watch.

How good are you at counting exactly 20 seconds?

*   Set the stop watch to zero.

*   Hold it so you cannot see the face.

*   Start the watch as you begin to count.

*   Stop the watch as soon as you finish.

*   Look at the stop watch. Write down the reading.

*   Do this four more times.

Did you get better with practice?

What is the mean or average of your five readings?

Now test your friends. On average, are they better than you?

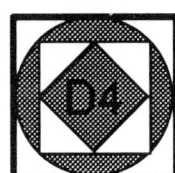

1 This graph tells the story of bath time for two muddy puppies.
Finish the story.

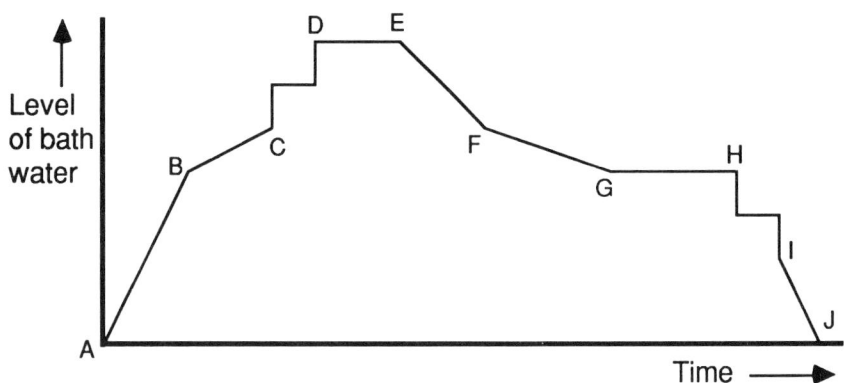

A to B: hot and cold water both go into the bath.

B to C: cold water only goes into the bath.

C to D: the two puppies get into the bath.

D to E: one puppy is washed.

E to F: the other puppy knocks out the plug.

F to G: the puppies splash water over the floor and walls.

G to H: ....

H to I: ....

I to J: ....

2 You need some writing paper. Make up a story to go with this graph.

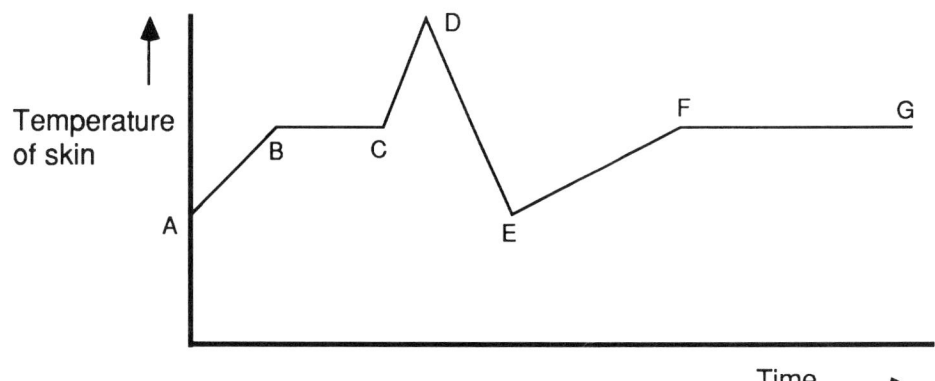

3 You need some paper and a ruler.

Draw a graph to illustrate the growth of a runner bean.

Write a story to go with it.

Data 6

1 This decision tree has sorted the numbers from ① to ⑯ .

The decision boxes labelled A to P have questions in them.

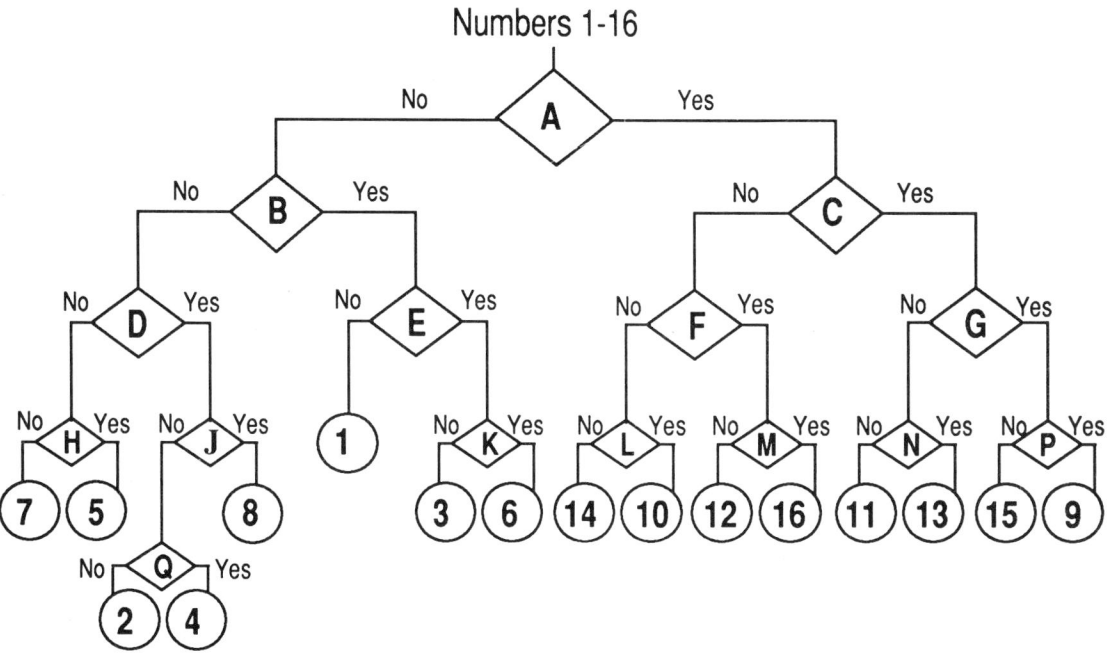

Choose one of the questions below for each decision box.

Use each question at least once.

Write the box letters by the questions.

Is it odd?	Is it half a dozen?
Is it even?	Is it unlucky?
Is it a multiple of 3?	Is it a triangular number?
Is it a multiple of 4?	Is it a square number?
Is it a multiple of 5?	Is it a cube number?
Is it greater than 8?	

2 Imagine you have a rubber, a screw, a jam jar, a paper clip, a piece of card, a pencil lead, a pebble and a paper tissue.

What different ways are there of sorting the objects?

Show your results on decision tree diagrams.

1 You need a 30cm ruler and some squared paper.

How quickly do you react? Work with a friend.

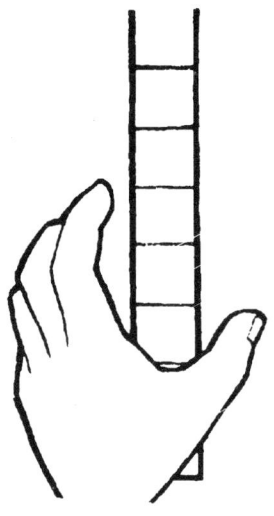

- Hold the ruler at the 30 cm mark.
- Your friend should put a finger and thumb near to the lower end.
- Drop the ruler for your friend to catch.
- Try three times with the writing hand and three times with the other hand.
- Take the best result for each hand.
- Now ask other friends to try.

Record the results on a tally chart like this . . .

Catching distance	1-10cm	11-20cm	21-30cm
Writing hand			
Non-writing hand			

. . . and also on this one.

Catching distance	1-5cm	6-10cm	11-15cm	16-20cm	21-25cm	26-30cm
Writing hand						
Non-writing hand						

Which chart was the most useful? What did it tell you?

2 You need pencil and paper and a tape measure marked in cm.

Measure round the wrist and the neck of each person in the class.

Make some charts to show your results grouped in different ways.

What did you discover?

1 You need some squared paper and a metre stick or tape measure.

Choose a distance that you don't know: for example, the width of the school gate, or the height of a tall cupboard.

Ask everyone in the class to estimate your chosen length.

Don't make your own estimate yet.

Keep a record of the results.

Arrange the estimates in order.

Make a graph to show the results.

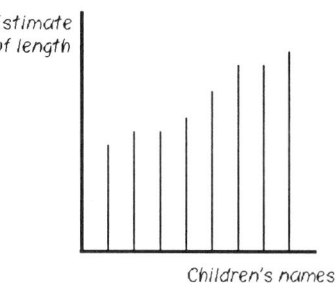

a. Which estimate is greatest?

b. Which is least?

c. What is the range of the estimates?

d. What is the middle estimate?

e. Now make your own estimate. What is it? Why?

f. Now go and measure the length. What is it?

g. Is this a good way to estimate lengths or distances?

2

Estimate the total number of letters in this square.

Find different ways of doing it.

Are they arranged evenly?

What about different letters, different sizes, ..?

Draw some graphs to show your results.

You need Data 9B. Organise the information to help you discover more about the people buried in the churchyard.

1 **Period**

a. When was earliest birth?　b.　When was the latest death?

c. In what years did most burials take place?

2 **Names**

a. How many different surnames were there?

b. What was the most common surname?

c. What was the most common first name for men?

d. What was the most common first name for women?

e. What are the most common first names now?

3 **Burials**

a. What was the total number of burials?

b. What percentage were males?

c. What percentage were females?

d. Plot a pie chart showing the results.
   Are they what you would have expected?

4 **Life expectancy**

a. Plot a frequency graph showing how old people were when they died. At about what age did most people die?

b. Plot a scattergraph to compare ages at death for males and females. Did men or women live longer?
   Can you think of a reason?

c. What was the probability that you would die before 60 if you were a man? Or a woman?

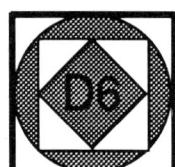

The information is from gravestones in Great Bedwyn churchyard.

Great Bedwyn is a very small market town in Wiltshire.

In 1850, there were about 2000 people living in the area.

SURNAME	FORENAME	SEX	AGE	DATE
TUCKER	JOHN	M	53	1863
HUMPHRIES	GEORGE	M	59	1858
HUMPHRIES	AMY	F	75	1869
EDWARDS	ELIZABETH	F	82	1870
THOMAS	ANTHONY	M	22	1855
BARTHOLOMEW	ISAAC	M	68	1870
BENSKIN	CHARLES	M	66	1857
MAY	PHOEBE	F	55	1863
MAY	JOSHUA	M	20	1861
FOXELL	DINAH	F	69	1865
MAY	JOHN	M	74	1857
PEATY	THOMAS	M	28	1871
WARD	ANN	F	60	1874
SMALLBONES	JOHN	M	76	1862
COOK	GEORGE	M	85	1859
POCOCK	SAMUEL	M	9 months	1873
DOBSON	ELIZABETH	F	66	1874
NEALE	MERVYN	M	6 months	1870
EDWARDS	CHARLES	M	63	1866
NEALE	ANN	F	63	1859
CARTER	ZABULON	M	34	1869
KINGSTON	ISAAC	M	73	1867
SMALLBONES	HANNAH	F	45	1871
CLARK	WILLIAM	M	80	1862
GARLICK	ISAAC	M	63	1860
THATCHER	MARY	F	70	1859
KNAPP	GEORGE	M	50	1875
SAWYER	SARAH	F	61	1868
PIKE	MARIA	F	81	1855
SYMONDS	SELINA	F	24	1872
GRANT	ELIZABETH	F	29	1872
MAY	THOMAS	M	69	1857

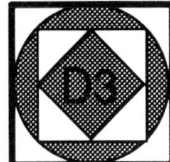

Probability 1

1 These words are used to describe how probable things are.

| very likely | fifty-fifty | probable | improbable | impossible | certain | unlikely |

Write all the words in order. Start with the lowest probability.

How likely are these? Write a word next to each.

    a. **I will watch television tonight.**

    b. **It will snow next Christmas.**

    c. **I will grow taller than my mother.**

    d. **I will get to school on time next Monday.**

    e. **I will see a horse and cart on my way home.**

    f. **I will buy a new tape next month.**

Write some statements of your own. Say how likely each one is.

2 You need some counters and a dice.

Play this game with a friend.

Rules

- Each player starts with 9 counters.

- Take turns to roll the dice.

- The first player wins *odds*. The second player wins *evens*.

- If 1 or 3 or 5 is rolled, *evens* gives *odds* that number of counters.

- If 2 or 4 or 6 is rolled, *odds* gives *evens* that number of counters.

- The winner is the first to gain all the counters.

Play the game five times. Record your results.

How many games did odds win?

How many did evens win?

Is this a fair game? Give your reasons.

1 There are four different pairs of
socks on the line. If it is dark,
how many socks must be taken to
be certain of at least one matching pair?

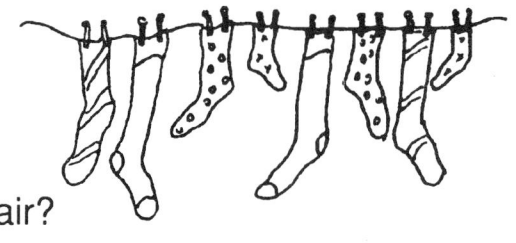

2 You need a bag, and 4 different coloured cubes.

Guesses			
1st	2nd	3rd	4th

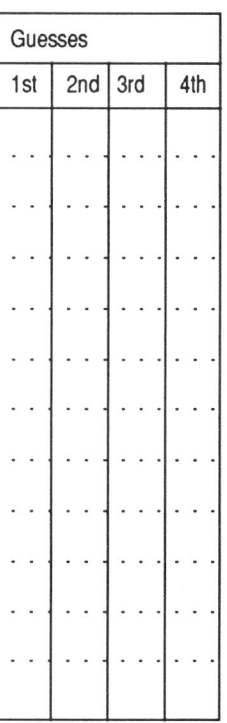

Put the four cubes in the bag and shake it.

Without looking, take a cube from the bag.

Before you do so, guess its colour.

If you are right, put a tick in the 1st column.

If you are wrong, put a cross.

Put the cube on the table.

Carry on until you have taken out all four cubes.

Do this experiment 12 times.

What is the probability of being correct on the
first guess?　　　On the last guess?

3 You need three boxes, three red cubes and three blue cubes.
Make three labels: RR for red red, BB for blue blue, RB for red blue.

Ask a friend to put two red cubes in one box, two blue cubes in
another box, and one red and one blue cube in the third box.
Your friend should put the **wrong** label against each box.

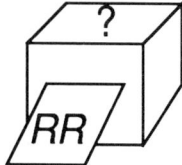

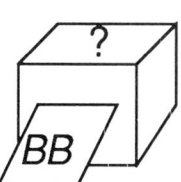

Take one cube from any box, look at its colour, and put it back.
How many times must you do this before you can say correctly
what is in each box?

Probability 3

1 Mr Jones cooked some pies.

He made pies in three different sizes, with four kinds of filling.

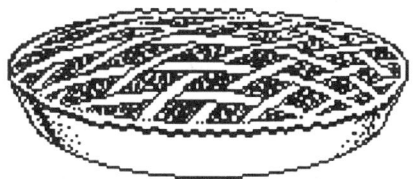

How many different pies could he have cooked?

2 Twins can be identical or non-identical.

With triplets, there are three possibilities:

all three identical

two identical and one not

all three non-identical

List all the possibilities for quins. What about sextuplets?

3 Five goals were scored in a hockey match. The final result was 2-3.
Write the possible scores as each goal was scored.

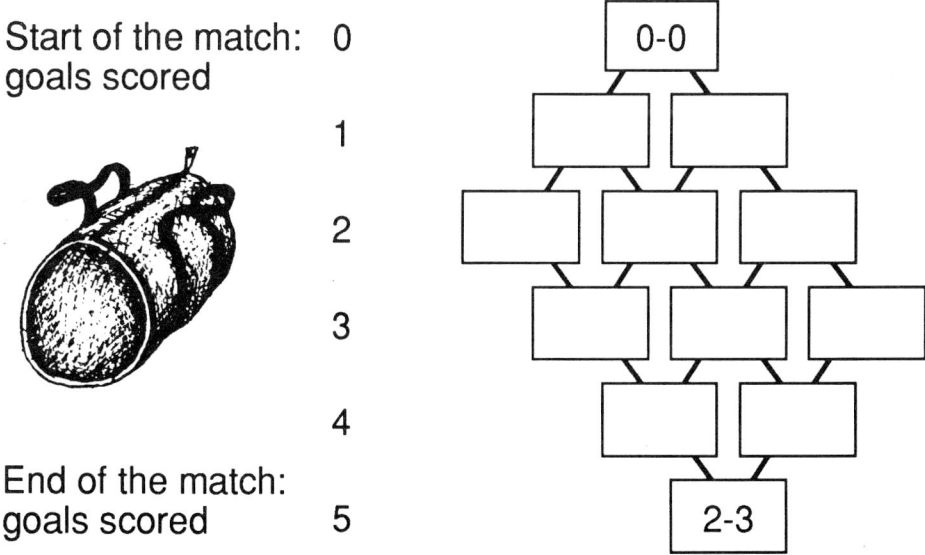

Start of the match: 0
goals scored

1

2

3

4

End of the match:
goals scored   5

0-0

2-3

In how many different ways could the final score be reached?

Investigate for other final scores.

You need some coloured pens and suitable paper.

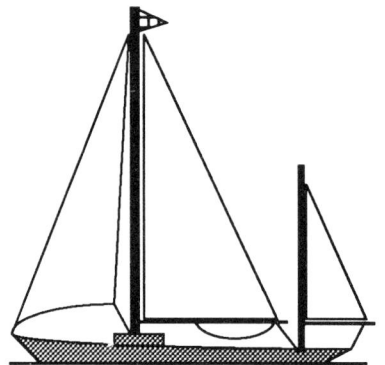

1 Chris drew a boat with three sails.
He had three different coloured pens.
He used a different colour for each sail.

    a. How many different ways could he colour the two tall sails?

    b. How many different ways could he colour all three sails?

Chris found a fourth coloured pen.

    c. How many ways could he now colour the two tall sails?

    d. How many ways could he colour all three sails?

2 Saima drew a train with three carriages and an engine.

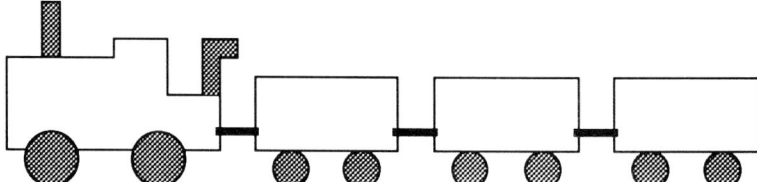

She had two coloured pens.
She shaded each part of the train with one colour.

    a. How many different ways could she colour the engine and the first carriage?

    b. How many different ways could she colour the whole train?

   What if Saima had three coloured pens?

3 Two squares on a 3 x 3 grid are coloured at random.
What is the probability that they touch edge to edge?

What about a 4 x 4 grid?

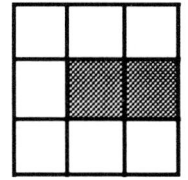

1 You need two dice and some squared paper for this experiment.

Roll both the dice.
Add the two scores.
Record the total by shading a square in the correct column.
Do this 60 times.

Which total occurred the most frequently?
Is this what you expected?

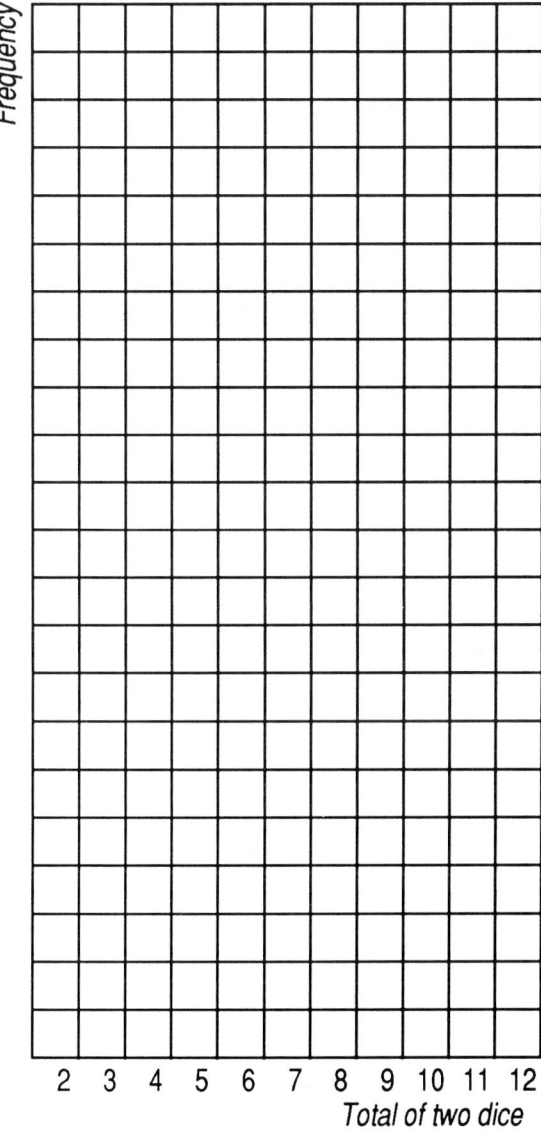

2 Complete a number square to show the possible total scores when rolling two dice.

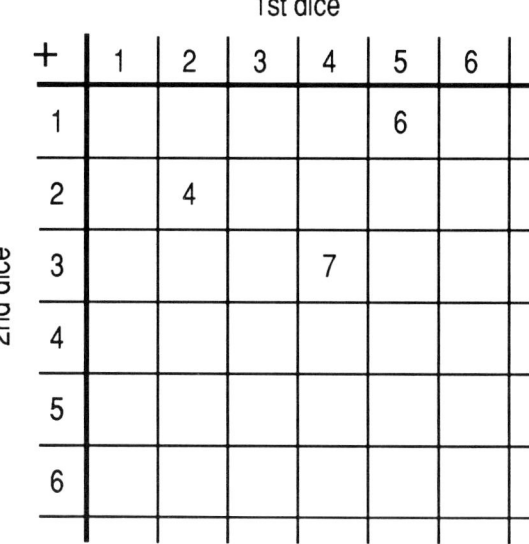

1st dice

+	1	2	3	4	5	6
1					6	
2		4				
3				7		
4						
5						
6						

2nd dice

Which number occured most frequently in the square?
Is this the same as in your experiment?

3 With two dice, what is the probability of a total score of ten?
If you rolled the dice 60 times, how often would you expect the total score to be ten?

1 You need two dice and some counters.

Play this game with a friend.

Rules

- Take turns to roll both the dice and add the two scores.
- If the sum is 2, 3, 4, 5, 10, 11 or 12, the first player takes a counter.
- If the sum is 6, 7, 8 or 9, the second player takes a counter.
- The winner is the first to gain five counters.

Play the game several times.

How often did each player win?

Is this a fair game?   Write your reasons.

2 You need Probability 6B, two dice and two coins.

Play the **Obstacle Race** with a friend.

At each stage, players choose one from a pair of two obstacles.

To get past an obstacle, follow the instructions in the box.

Both players are allowed to practise.

Do some experiments to help you choose the easier obstacles.

Rules

- Each player chooses a route for the course.
- Players take turns to roll the dice or toss the coins, according to the instructions in the boxes.
- Players who are successful can move on to the next obstacle.
- Players who are unsuccessful must try again on their next turn.
- The winner is the first player to reach the finish.

What is the best route to choose for the race?

Write your reasons.

3 Make up and investigate some more obstacle races.

Probability 6B

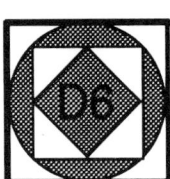

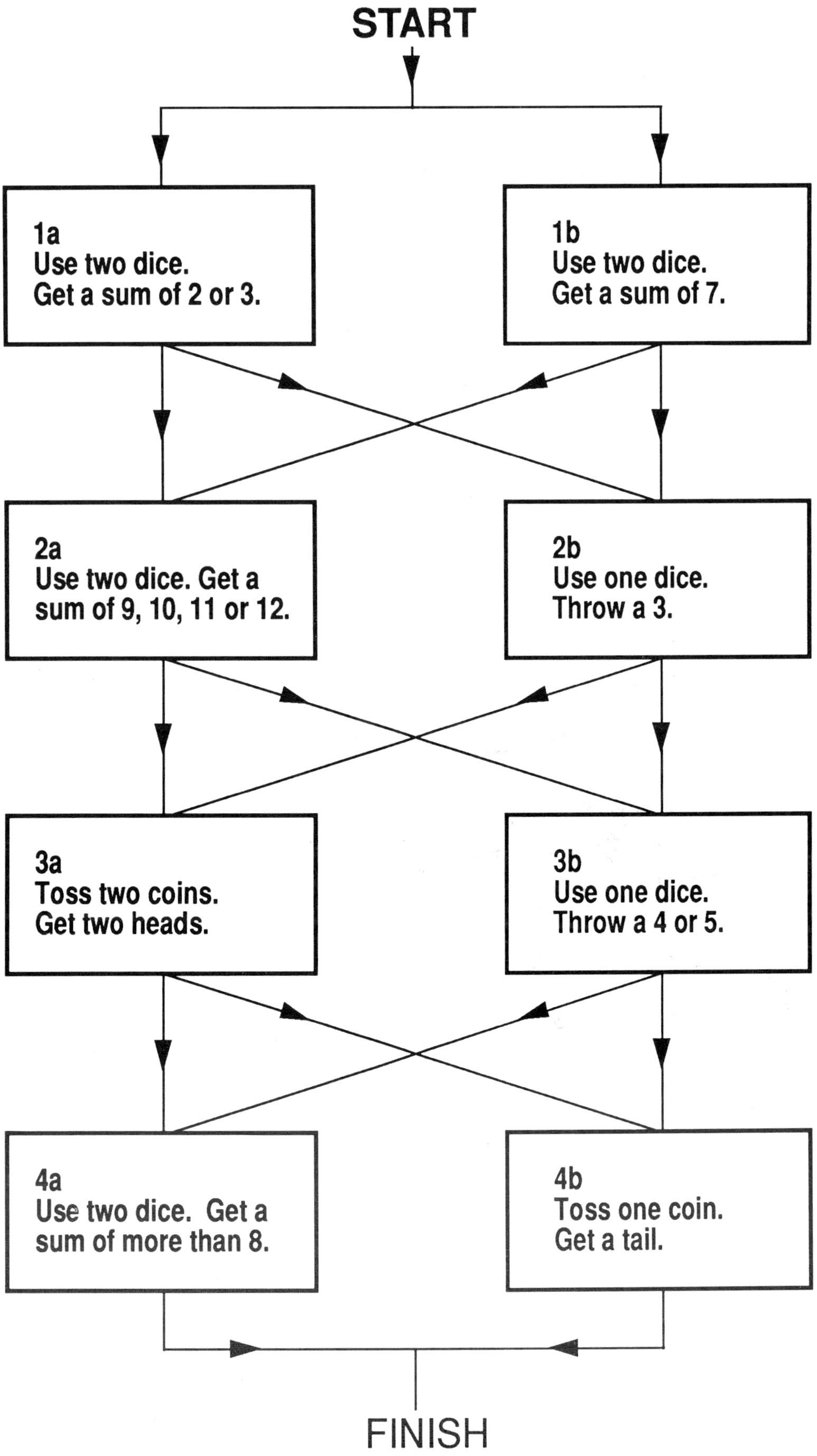
START
1a
Use two dice.
Get a sum of 2 or 3.
1b
Use two dice.
Get a sum of 7.
2a
Use two dice. Get a
sum of 9, 10, 11 or 12.
2b
Use one dice.
Throw a 3.
3a
Toss two coins.
Get two heads.
3b
Use one dice.
Throw a 4 or 5.
4a
Use two dice. Get a
sum of more than 8.
4b
Toss one coin.
Get a tail.
FINISH

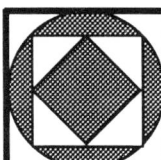

Cut round these.

0	1	2	3
4	5	6	7
8	9	+	—
×	÷	=	

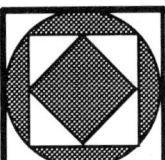

Solutions

Number puzzles 1 page 5

1

```
6 4 2    2 3 8
5   7    6   4
1 8 3    5 7 1
  12       13

7 6 1    3 5 7
3   5    4   2
4 2 8    8 1 6
  14       15
```

2   7 in line.

```
4 2 1   3 2 2   2 3 2   1 3 3   3 2 2
2   4   2   4   3   3   3   3   3   3
1 4 2   2 4 1   2 3 2   3 3 1   1 4 2

5 2 1   4 2 2   3 2 3   2 1 5   3 1 4
2   2   2   2   2   2   2   2   2   2
1 2 5   2 2 4   3 2 3   4 3 1   3 3 2
```

8 in line.

```
6 1 1   5 1 2   4 1 3   3 1 4
1   3   1   3   1   3   1   3
1 3 4   2 3 3   3 3 2   4 3 1
```

Number puzzles 2 page 6

1 There are various possibilities. Encourage finding as many as possible:

eg 3-2=1; 2+2=4; 3+2=5; 2x3=6; 8+2=10; 8+3=11; 2x8=16; 3x8=24.

You could also make 9=3x3, 64=8x8, or four digit numbers such as 2338.
The use of the constant may be possible: eg 2 x = = may give 8; 8 x = = may give 512.
Some children may be interested in making negative numbers.

Using 1, 3 and 9 you can make 1, 2, 3, 4, 6, 8, 9, 10, 12, 18, 27, or 81, as well as other numbers produced in the various ways described above.

2 Tom has 19 books, or 39, 59, 79 . . .

Number puzzles 3 page 7

1 7, 3, 4, 6.

2 12 rubbers balance 9 cubes.
The red parcel is probably heavier.

3 8

Number puzzles 4 page 8

1 With 1,2,3,4,5,6: four distinct solutions.

With 1,2,3,5,6,7: two solutions

With 1,2,3,4,6,7: two solutions.

2 The differences add to 26. A number of arrangements are possible: eg

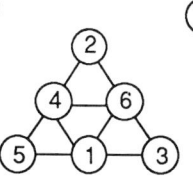

Number puzzles 5 page 9

1 Consecutive sums can be made in various ways: for example, 39 = 19 + 20, or 12 + 13 + 14.
Numbers which are divisible by an odd number, like 3 or 5, can be expressed as the sum of 3 or 5 numbers: for example, 30 = 3 x 10 = 9 + 10 + 11, or 30 = 5 x 6 = 4 + 5 + 6 + 7 + 8.
1000 = 198 + 199 + 200 + 201 + 202.
These numbers cannot be expressed as consecutive sums: 2, 4, 8, 16, 32, . . .

2 Using addition of 3s and/or 5s, the only numbers greater than 3 which cannot be made are 4 and 7.
Using addition of 4s and/or 7s, it is possible to make 4, 7, 8, 11, 12, 14, 15 and 16, and from 18 onwards. The largest number which cannot be made is 17.

3 For example: 12 + 3 - 4 + 5 + 67 + 8 + 9 = 100   123 + 45 - 67 + 8 - 9 = 100
                  123 - 4 - 5 - 6 - 7 + 8 - 9 = 100   123 - 45 - 67 + 89 = 100

Number puzzles 6 page 10

1 eg:
```
  541        543        134        123
  +32        -12        +25        -54
  ───        ───        ───        ───
  573        531        159         69
```

2
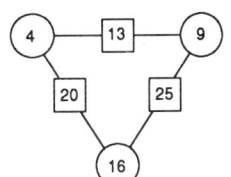

3 25

Number puzzles 7 page 11

1

13	8	12	1
3	10	6	15
2	11	7	14
16	5	9	4

Change
5 and 6,
7 and 8,
and 1 and 4.

2

7	10
5	13

3 For 67: 12,17 or 11,27.
For 111:
11,67; 12,57; 13,47;
21,48; 22,38; 31,29;
(rotate for 6 more pairs).

Number puzzles 8 page 12

1
```
a.   46      b.   38      c.   33
   + 28         - 29         + 87
   ────        ────        ─────
     74            9          120
```

3 Rods can be placed end to end, or side by
side. With 5s and 6s, all lengths up to
40cm can be found.   With 5s and 7s, all
but 1cm can be measured.
If rods are end to end only, with 5s and 7s
it is impossible to make 1, 2, 3, 4, 6, 8, 9,
11, 13, 16, 18 and 23.

2
```
43 - 37 =  6        17 + 28 = 45
37 - 29 =  8        28 + 37 = 65
43 + 37 = 80        28 - 17 = 11
37 - 17 = 20        43 - 17 = 26
37 - 28 =  9        28 + 43 = 71
```

```
 1 = 6 - 5            13 = (3 x 6) - 5
 3 = (3 x 5) - (2 x 6)   14 = (4 x 5) - 6
 9 = (3 x 5) - 6         16 = (2 x 5) + 6
11 = 6 + 5            17 = (2 x 6) + 5
```

Number puzzles 9 page 13

1 36 + 19 + 11 + 7 and 49 + 23

2
```
M   A   G   I   C
12  14  11  8   10
```

3 The smallest difference is 1.
67 + 31 + 4 + 5 = 107
46 + 24 + 19 + 17 = 106

Number puzzles 10 page 14

1 The three readings added together give double the weight of all three.
The dog is 15kg; the girl is 32kg; the boy is 50kg.

2

London	9:00	9:40
Wimbledon	9:15	9:55
Woking	9:25	10:05
Basingstoke	9:55	10:35

Basingstoke	10:10	10:50
Woking	10:40	11:20
Wimbledon	10:50	11:30
London	11:05	11:45

London to Wimbledon is 15 minutes. Woking to Basingstoke is half an hour.

Number puzzles 11 page 15

1 15 + 1 = 16;
15 + 11 + 11 = 37;
51 + 15 + 11 + 11 = 88;
511 + 111 + 15 + 1 = 638. Other
solutions are possible.

2 Possibilities are:
325 + 275 + 170 + 230
117 + 483 + 262 + 138
81 + 414 + 275 + 230, etc.

3 For example, 128 + 574 - 693 = 9.

Number puzzles 12 page 16

1 2 with 9 apples
7 with 2 apples.
Total: 9 baskets.

2 54 x 3 = 162

3 40cm or 184cm.

4
1	2x3-4-1
2	1+2+3-4
3	2x3+1-4
4	1+2+4-3
5	12-3-4
6	1+3+4-2
7	24/3-1
8	2+3+4-1
9	23-14
Possibilities: 10	1+2+3+4

11	12+3-4
12	2x4+3+1
13	12+4-3
14	21-3-4
15	13+4-2
16	34/2-1
17	1x34/2
18	32-14
19	13+4+2
20	21+3-4

21	2x4+13
22	34-12
23	4x3x2-1
24	4x3x2x1
25	4x3x2+1
26	24+3-1
27	32-4-1
28	23+4+1
29	42-13
30	13x2+4

31	43-12
32	12x3-4
33	132/4
34	(14+3)x2
35	32+4-1
36	4x1+32
37	24+13
38	42-3-1
39	4x2+31
40	12x3+4

Number puzzles 13 page 17

1 36 = 36 x 1 or 18 x 2 or
12 x 3 or 9 x 4 or 6 x 6.

2 Cross out the first 3 in the 2nd row,
and the 6 in the third row.

3 8 windows: RRR, RRY, RYR, RYY, YRR,
YRY, YYR, YYY.
With 3 colours: 27 windows.
With 4 colours: 64 windows.

Number puzzles 14 page 18

1
x	5	4	9	2
2	10	8	18	4
3	15	12	27	6
7	35	28	63	14
1	5	4	9	2

2
13 x 14 = 182
9 x 10 = 90
14 x 15 = 210
18 x 19 = 342
27 x 28 = 756
43 x 44 = 1892
57 x 58 = 3306

3 Jake has 4
rabbits and
36 geese.

Number puzzles 15 page 19

1
a. 12, 24, 36 and 48 are four times the sum of their digits.

b. 45 is five times and 54 is six times the sum of their digits.
21, 42, 63 and 84 are seven times the sum of their digits.
72 is eight times and 81 is nine times the sum of their digits.

c. All the numbers are multiples of three.

2
a. 33, 27, 21, 15, 9 miles.
b. 108 miles.

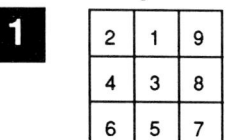

3 Octopus: 3 Mermaid: 7 Dolphin: 6

Number puzzles 16 page 20

1
2	1	9
4	3	8
6	5	7

Alternatively,
192, 273 or 327
in the top row.

2 Since 37 x 3 = 111, 37 x 3 x a = aaa,
where a is a digit.

3 Rose has 12 cows and 23 chickens.

Number puzzles 17 page 21

1 16.
11,111 - 5
= 11,106.

2 A=7, E=1,
G=5, L=9,
S=2, Y=6.

```
  976
+ 576
------
 1552
```

3
```
 1503    1035    1089
- 829   - 246   - 724
-----   -----   -----
 674     789     365
```

Other solutions are possible.

Number puzzles 18 page 22

1 A=4, B=8, C=9, D=1, E=3, F=6, G=5.   DC x EF = FBA.

2 56 ÷ 7 = 8; 81 ÷ 9 = 9; 546 ÷ 6 = 91 or 516 ÷ 6 = 86 or 576 ÷ 6 = 96;   469 ÷ 7 = 67;
40707 ÷ 9 = 4523 or 49707 ÷ 9 = 5523.

3 4, 9, 25 and 49 each have exactly 3 factors. 72 and 96 each have 12 factors.

Number puzzles 19 page 23

1 Nine spheres.

2 8, 10, 12 and 18kg or 6, 12, 14 and 16kg.

3 Balance nine against nine. If they balance, the heavier ball is one of the nine left on the table. Otherwise, the heavier ball is one of the nine in the pan which goes down. Take the set of nine with the heavier ball. Balance three against three to find the set of three with the heavier ball. Then balance one against one.

Number puzzles 20 page 24

1 Karen: one 40 and five 16s;
Tracey: two 23s and four 16s;
Dean: four 17s and two 16s.

2 $55 \times 56 = 3080$
$7 \times 29 \times 37 = 7511$
$69 \times 71 \times 73 = 357\ 627$

3 1st: 1336;
2nd: 1314;
3rd: 1306;
4th: 1263.

Number puzzles 21 page 25

1 There are 8 possible combinations of stamps so the maximum cost is £52.80.

$5 \times 18p + 19 \times 30p$; $15 \times 18p + 13 \times 30p$; $25 \times 18p + 7 \times 30p$; $35 \times 18p + 1 \times 30p$;
$10 \times 18p + 16 \times 30p$; $20 \times 18p + 10 \times 30p$; $30 \times 18p + 4 \times 30p$; $22 \times 30p$.

2 a. $4 \times 1738 = 6952$, $4 \times 1963 = 7852$.

b. $483 \times 12 = 5796$; $297 \times 18 = 5346$; $198 \times 27 = 5346$; $186 \times 39 = 7254$;
$159 \times 48 = 7632$; $157 \times 28 = 4396$; $138 \times 42 = 5796$.

c. For example: $158 \times 23 = 79 \times 46$ $174 \times 32 = 96 \times 58$ is the largest.

3 a. $27\ 548 \times 3 = 82\ 644$ b. $172 \times 145 = 24\ 940$

Number puzzles 22 page 26

1 Deer: 20 yrs. Elk: 30 yrs. Dormouse: 4 yrs. Chamois: 25 yrs. Porcupine: 15 yrs.

2 $16 \times 43 = 688$ $73 \times 57 = 4161$
$37 \times 43 = 1591$ $14 \times 29 = 406$

3 91kg.

Number puzzles 23 page 27

2 If Lisa takes 2 time units over each section, Jenny takes 1, 2 and 4 time units for the three sections. Jenny takes one time unit longer so Lisa wins by one mile.

Number puzzles 24 page 28

1 Largest: $521 \times 43 = 22\ 403$
Smallest: $123 \div 54 = 2.2777...$

3 $3.1622776...$

4 2.7 and 7.3.

2 Your calculator may do better than this.
8 keys . 1 + . 0 1 = = 0.12
10 keys 1 + 1 + . 1 = = = = 2.4
12 keys . 1 1 + = = = = = = = = = 0.88
9 keys 1 + . 0 1 = = = = 1.04
8 keys 1 . 1 + 1 . 1 1 = 2.21

Number puzzles 25 page 29

1 a. Sweden has the highest proportion of refugees: 1 in every 63 people.
b. The UK has the lowest proportion of refugees: 1 in every 525 people.

3 a. 32 great great great grandparents.
b. 1000 years is about 40 generations, so about 1 099 512 million ancestors.
The total population of the world in 900 AD is estimated at 350 million.

Number puzzles 26 page 30

1 9 128 376 is divisible by 8.
882 351 is divisible by 9.
603 482 is divisible by 11.
2 723 625 is divisible by 225.

2 eg 321 654; 3 816 547 290.

3 13 and 11. 7 and 13.

4 Any number of the form a^4b^2 has exactly fifteen divisors. The smallest such number is 144.

Money 1 page 31

1 1p, 2p, 3p, 5p, 6p, 7p, 8p, 10p, 11p, 12p, 13p, 15p, 16p, 17p, 18p.

2 50p can be made in 13 different ways using silver coins:
(1 x 50p);   (2 x 20p) + (1 x 10p);   (2 x 20p) + (2 x 5p);   (10 x 5p)
(1 x 20p) + (3 x 10p);   (1 x 20p) + (2 x 10p) + (2 x 5p);
(1 x 20p) + (1 x 10p) + (4 x 5p);   (1 x 20p) + (6 x 5p);
(5 x 10p);   (4 x 10p) + (2 x 5p);   (3 x 10p) + (4 x 5p);   (2 x 10p) + (6 x 5p);   (1 x 10p) + (8 x 5p).

3 Each child received three 20p and one 5p coin, or one 50p and three 5p coins.

Money 2 page 32

1 38p requires 5 coins.
77p requires 4 coins.
91p requires 4 coins.
44p requires 4 coins.

2 The parcel could cost:
15p, 20p, 30p, 35p,
40p, 50p, 55p, 70p.

3 Four 20p stamps.
Twelve 10p stamps.

Money 3 page 33

1 You can pay any amount up to 88p, with these exceptions:
4p, 9p, 14p, 19p, 24p, 29p, 34p, 39p, 40p to 49p, 54p, 59p, 64p, 69p, 74p, 79p, and 84p.
There are therefore 63 different amounts that can be paid.
An extra 2p coin would make a lot of difference!

2 You could send 15 different parcels:
£1.05, 90p, 85p, 75p, 70p, 65p, 60p, 55p, 50p, 45p, 40p, 35p, 30p, 20p, 15p.

3 The largest sum in coins you could have without being able to pay exactly £1 would be
£1.43: one 50p and four 20p coins, one 5p and four 2p coins.

4 Maria bought five 4p stamps, thirty 2p stamps and four 5p stamps.

Money 4 page 34

1 I got ten apples for £2,
including the two that were free.

3 Greatest value of stamps: 50p

4	3	5	2
5	2	1	4
1	4	3	5
3	5	2	1

2 Amounts up to £1 which can be paid exactly
using one or two coins are:
1p, 2p, 3p, 4p, 5p, 6p, 7p, 10p, 11p, 12p, 15p,
20p, 21p, 22p, 25p, 30p, 40p, 50p, 51p, 52p,
55p, 60p and 70p.
All others need at least three coins.
Amounts which need one, two or three coins
are those listed above, plus:
8p, 9p, 13p, 14p, 16p, 17p, 23p, 24p, 26p, 27p,
31p, 32p, 35p, 41p, 42p, 45p, 53p, 54p, 56p,
57p, 61p, 62p, 65p, 71p, 72p, 75p, 80p, 90p.
All others need at least four coins.

Money 5 page 35

1
a.   Tennis ball: £2.00
b.   Football: £2.40

3
Lisa: 3 at 5p and 15 at 3p;
Pat: 6 at 5p and 10 at 3p;
Jacky: 9 at 5p and 5 at 3p.

a.   Lisa bought the most.
b.   Jacky bought the fewest.

2 There are 6 possible combinations.

£4.80	£3.40	£1.80
£4.80	£2.70	£1.80
£3.90	£3.40	£2.70
£3.90	£3.40	£1.80
£3.90	£2.70	£1.80
£3.40	£2.70	£1.80

If more than one bear of the same size can be
bought, there are 11 more possibilities.

Money 6 page 36

1 Ten large eggs @ 50p, ten medium eggs @ 10p, 80 small eggs @ 5p.

2 Ann's brother is Nick; he spent £3. Mary's brother is Simon; he spent £8.
Jane's brother is Andrew; he spent £3. Kate's brother is Tim; he spent £8.
Altogether, the girls spent £10 and the boys spent £22. Total: £32.

3 Andover to London is £8; London to Cambridge is £12. Total: £20.

Negative numbers page 37

1

	Final score	Strokes		Final score	Strokes
Andrews	-11	277	Fielder	2	290
Brown	-13	275	Green	-1	287
Cox	-11	277	Howard	1	289
Dobson	-5	283	Innes	6	294
Evans	-1	287	Jones	4	292

2 Dobson: 72 strokes, won 9 holes; Howard 74 strokes, won 6 holes.

Fractions 1 page 38

1

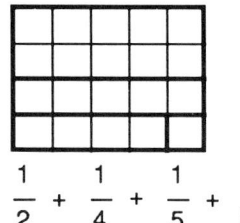

$$\frac{1}{2} + \frac{1}{4} + \frac{1}{5} + \frac{1}{20}$$

$$\frac{1}{2} + \frac{1}{3} + \frac{1}{6}$$

$$\frac{1}{2} + \frac{1}{5} + \frac{1}{6} + \frac{1}{10} + \frac{1}{30}$$

2 60 sweets in the bag.

3 Eg, $\frac{7}{24}$. There are many others.

Fractions 2 page 39

1 Adding 1 to each of the numerator and denominator increases the fraction if the original fraction < 1, and decreases it if the original fraction > 1.
The opposite is true if 1 is subtracted, rather than added.

2 a. $\frac{4}{15}$

   b. 18

3 Eg:

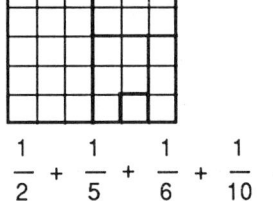

4 eg $\frac{5}{12}$

Fractions 3 page 40

1 18.

2 Parveen: 18 years old   Afzal: 14 years old

3 There are six different paths passing through the minimum number of cells, which is five.
The least total for a path is $2\frac{7}{8}$.

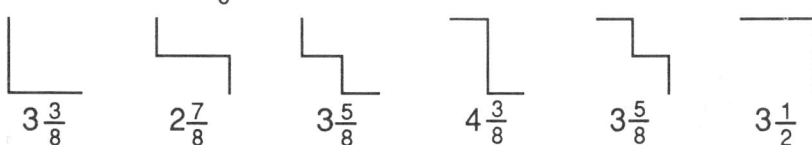

$3\frac{3}{8}$ $2\frac{7}{8}$ $3\frac{5}{8}$ $4\frac{3}{8}$ $3\frac{5}{8}$ $3\frac{1}{2}$

The maximum total of $6\frac{1}{8}$ is from the path which passes through all nine cells.
By adding then subtracting alternately, maximum is $1\frac{1}{8}$ and minimum $-1\frac{1}{4}$.

Fractions 4 page 41

1 $\frac{1}{49}$ and $\frac{1}{87}$

2

$\frac{1}{7}$	0.1428571
$\frac{2}{7}$	0.2857142
$\frac{3}{7}$	0.4285714
$\frac{4}{7}$	0.5714285
$\frac{5}{7}$	0.7142857
$\frac{6}{7}$	0.8571428

3 Digits after decimal point are always 2 x previous number with the decimal point ignored.

The first six digits are the same but in a different order.

1.1428571
1.2857142
2.2857142

With 12 digits, one seventh is 0.142857142857

Percentages page 42

1 No difference.

2 First cow cost £750. Second £500.
Total sale price was £1200.
The farmer made a loss of £50.

3 Since 1 x 1.5 x 1.2 = 1.8, the pay rise over two years was 80%, not 70%.

4 26 people out of the 41 asked said 'yes'.

Patterns 1 page 43

1 10,1; 9,2; 8,3; 7,4; 6,5.

2 20,2; 18,4; 16,6; 14,8; 12,10.

4

Number of steps	1	2	3	4	5	6	7	8	9	10
Number of squares	1	3	6	10	15	21	28	36	45	55

A staircase with 20 steps would have 210 squares.

The rule is n(n+1)/2.

Patterns 2 page 44

1 6, 9, 12, 15, 18, 21, 24, 27.
8, 12, 16, 20, 24, 28, 32, 36.
22, 20, 18, 16, 14, 12, 10, 8.

2 Eg 46 is not in the pattern of 3s since it is 4 more than 42 which is in the pattern.

The pattern of 3s slopes from left to right on grids with 2, 5, 8, 11, 14, 17 ... columns.

Patterns 3 page 45

1 The sequence is 1, 3, 5, 7, 9, 11, . .
The 20th number is 39.
The 1000th number is 1999.
The general rule is 2n - 1.

2 The sequence is 1, 4, 7, 10, 13, . .
The 20th number is 58.
The 100th number is 298.
The general rule is 3n - 2.

3 The smallest number which will divide exactly by 5, 7, 9 and 21 is 315.

Patterns 4 page 46

2 If the height and the width have a factor in common, then divide each side by that factor, to produce a smaller rectangle. For example, a 30 x 18 rectangle produces the same result as a 5 x 3 rectangle. Then:
 if both width and height are odd, the ball lands at C;
 if AB only is even, the ball lands at B. If AD only is even, the ball lands at D;
 if both sides are even, the ball lands in the same corner as with a smaller one.

Patterns 5 page 47

1

1	2	3	4	5	6	7	8	9	10
1	5	9	13	17	21	25	29	33	37

The general formula is 4n - 3.
The 20th number has 77 counters.
The 100th number has 397.

2

1	2	3	4	5	6	7	8	9	10
1	4	9	16	25	36	49	64	81	100

The general formula is n^2.
The 20th number has 400 counters.
The 100th number has 10 000.

3 For 7 people, 7 x 6 = 42 presents are needed; for 16 people, 16 x 15 = 240 are needed.

Patterns 6 page 48

1

Times folded	1	2	3	4	5	6	7
Fold lines	1	3	7	15	31	63	127
Regions	2	4	8	16	32	64	128

If the number of times folded is n, then the number of regions is 2^n and the number of fold lines $2^n - 1$. After ten folds, there are 1024 regions and 1023 fold lines.

2

1 2	2 1	3 3	4 2	5 4
6 1	4	3	7 3	6
8	8 4	3	9 2	10 4
11 4	12 9	13 1	2	0
8	14 6	4	15 5	6

d=7, e=8, f=12, g=11

Patterns 7 page 49

1 26. I must wait 38 years.

3

+	1^2	2^2	3^2	4^2	5^2	6^2	7^2
1^2	2	5	10	17	26	37	50
2^2	5	8	13	20	29	40	53
3^2	10	13	18	25	34	45	58
4^2	17	20	25	32	41	52	65
5^2	26	29	34	41	50	61	74
6^2	37	40	45	52	61	72	85
7^2	50	53	58	65	74	85	98

2 A cube with edge 7.937cm holds approximately half a litre.

4 Experimenting with possible last digits rules out numbers ending in 2, 3, 4, 7, 8 or 9. Simple examples are 0, 1, 5 and 6. Other cases are 25, 76, 376, 625.

1, 2, 4, 5, 7, 8, 10, 13, 15, 16, 20, 23, 25, 28, 31, 32, 37, 39, 40, 47, 52, 55, 58, 60, 63, 64, 71, 79, 80, 85, 87, 92 and 95 cannot be written as the sum of three squares. Alll can be written as the sum of four squares.

Patterns 8 page 50

1 **Solid cubes**

Size	2	3	4	5	6	7	8
Number	8	27	64	125	216	343	512

For a 20cm solid cube: 8000 cubes

3 **Skeleton cubes**
4 pillars plus 8 shorter lengths.
Sequence: 8, 20, 32, 44, 56, 68, 80 . . .
In general: 4n + 8(n-2) or 12n - 16.

2 **Hollow cubes**

Size	2	3	4	5	6	7	8
Number	8	26	56	98	152	218	296

For a 20cm hollow cube: 2168 cubes

In general: two full faces $2n^2$
plus four middle faces $4(n-1)(n-2)$
totalling $6n^2 - 12n + 8$.

Patterns 9 page 51

1 Encourage use of the patterns rather than putting out a long line of counters.
In the first pattern, the 65th counter would be blue. The 17th blue counter would be the 27th.
In the second, the 65th counter would be blue. The 15th red counter would be 35th in line.

2 32 can be made by 8 + 4 or 8 + 2 + 2.
40 can be made by 5 + 4 + 2 + 1 or 5 + 2 + 2 + 2 + 1.
Maximum product for 12 is 81, made from 3 + 3 + 3 + 3.
In general, maximum for 3n is 3^n, for 3n + 1 is $4 \times 3^{n-1}$, and for 3n + 2 is 2×3^n.

Patterns 10 page 52

1 The result will always be divisible by 11 for two or four digit numbers.

2 Every time the addition results in at least one 'carry' a further addition is required.

3 The result will always be 1089 unless the three digit number is a palindrome, when the first subtraction gives zero.

Patterns 11 page 53

1 221 pages. 42 fives.

3 48 children aged 8 years.

2 With four digits, the outcome is always 6174, which is known as Kaprekar's constant.
The longest chains require seven subtractions and start with these numbers:
1236, 1246, 1279, 1346, 1356, 1389, 1456, 2347, 2357, 2457, 2467, 2567, 3458, 3468, 3568, 3578, 3678, 4569, 4579, 4679, 4689, 4789.

Patterns 12 page 54

1 Eg, ◆ means *is the larger of*;
✳ means *double the first
number and add the second.*

4 There are five different chains. Shortest
is 5 - 5 - 0. Longest has 60 links.

With multiplication, then division by 7, one
chain has 3 links, one has 8, and one 23.

2 The ▲ means *to the power of*.

3 In the first sequence, ■ means *multiply by
3 and add 1*; ▼ means *subtract 2*. The
next numbers are 43, 41, 124, 122, ...

In the second sequence, ■ means *divide
by 2*; ▼ means *multiply by 3 and subtract 1*.
The next numbers are 28, 83, 41.5, 123.5, ..

Patterns 13 page 55

1

Pins on edge (B)	3	4	5	6	7
Area (A)	0.5	1	1.5	2	2.5

$B = 2A + 2$

2

Pins inside (P)	0	1	2	3	4
Area (A)	3	4	5	6	7

$P = A - 3$

These are special cases of Pick's theorem relating area (A) to the pins on the boundary (B)
and pins inside (P) by the formula $B = 2A + 2 - 2P$.   If $B = 10$, then $P = A - 4$.

Patterns 14 page 56

1 If c is the highest common factor of w and h, then $s = w + h - c$.
For example, if $w = 12$ and $h = 16$, then $c = 4$ and $s = 12 + 16 - 4 = 24$.

2
a. Quadrilateral: 2 Pentagon: 5 Hexagon: 9
b. For n sides, the total number of diagonals $n(n-3)/2$.
c. Pentagon: 2 Hexagon: 3 Heptagon: 4
d. For n sides, the number of non-crossing diagonals is $n - 3$.

Patterns 15 page 57

1 Sixth day: 1+6+12+18+24+30 = 91
18th day: 919
19th day: 1027

General formula is $1+3n(n-1)$

2

Length of path	1	2	3	4	5	6	7
Number of tilings	1	2	3	5	8	13	21

In general, each number of tilings equals
the sum of the two previous numbers.

3 The strip can be used to measure directly any whole number of inches from 1 to 36.
For example, 17 inches is the interval between the 3 and 20 inches marks.
A strip one foot long could be marked 1, 3, 7, 11; or 1, 2, 3, 8; or 1, 5, 9, 11.

Co-ordinates 2 page 59

1 (32,40) Two spaces east of target.

2 16 counters can be
placed like this but a
17th ends the game.

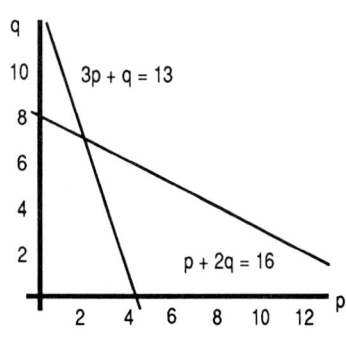

Co-ordinates 3 page 60

2 To the nearest half centimetre: a. 3.5cm   b. 6.5cm

Co-ordinates 5 page 62

1 For $p + 2q = 16$, for example:

p	0	2	4	6	8
q	8	7	6	5	4

For $3p + q = 13$, for example:

p	0	1	2	3	4
q	13	10	7	4	1

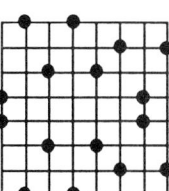

2 a. liquorice: 7p b. chew: 2p

3D shape 1 page 63

1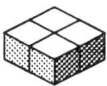

These two shapes are both half cubes.

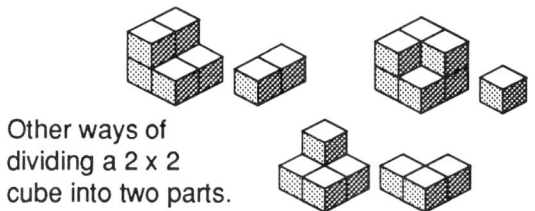

Other ways of dividing a 2 x 2 cube into two parts.

2 The solid cube is 4 x 4 x 4. It hides a 2 x 2 x 2 cube, so 56 cubes on outside.

3D shape 2 page 64

1 3 to complete the loop.
5 to get back to the start.

2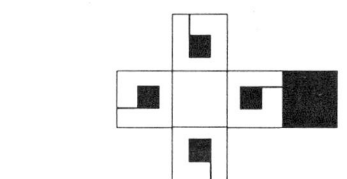

3 See solution to 2D shape 3 on page 123.

3D shape 3 page 65

1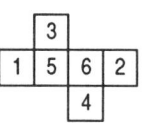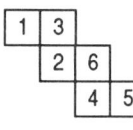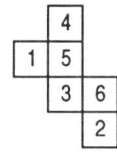

2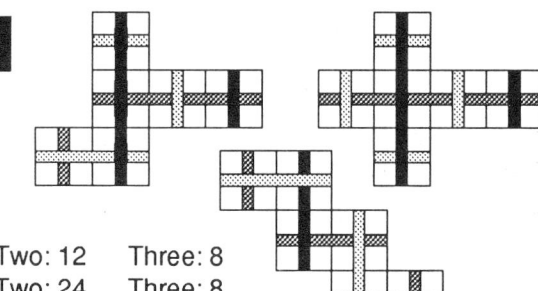

3
	None	One	Two	Three
For a 3 x 3 x 3 cube	1	6	12	8
For a 4 x 4 x 4 cube	8	24	24	8
For a 5 x 5 x 5 cube	27	54	36	8

3D shape 4 page 66

1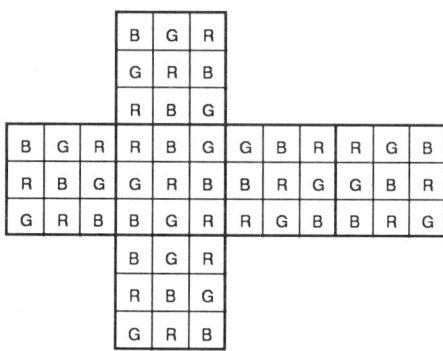

2 8 different shapes can be made.

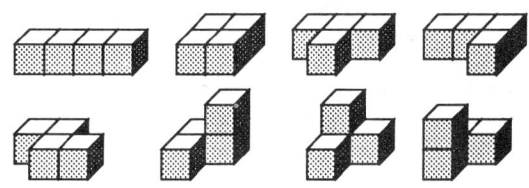

Outside surface area varies since in some of the shapes some of the cubes make contact by more than one face.

3D shape 5 page 67

1 Eight ways with four colours. When folded, A and B give two distinct tetrahedra. All the rest are rotations of A or B.

Eight ways with two colours. When folded, A, B, D, G and H give five distinct tetrahedra.

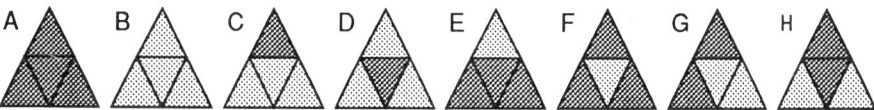

2 Any cube can be folded flat by cutting just seven edges.

3D shape 6 page 68

1 17 shapes.

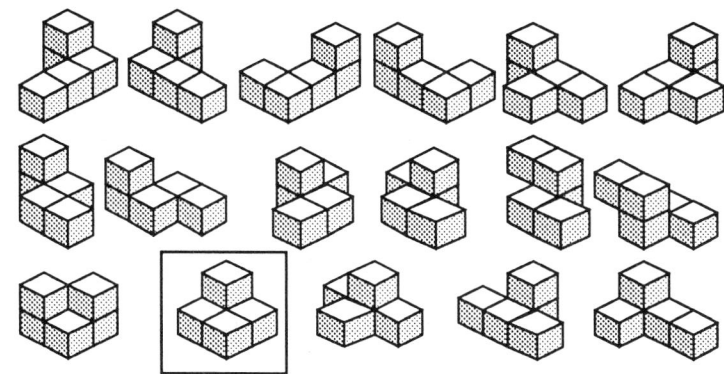

Minimum surface area is the boxed shape: 20 sq. units.

2 Paint each cube so that the three faces meeting at one corner are all red and the three faces meeting at the opposite corner are all blue.

3
a. At least 3 colours are needed to paint a cube so that adjacent faces differ.
b. With up to four colours there are 10 possibilities. The last four shown use only three.

G R Y R G B	G Y R Y G B	B R Y R B G	B Y R Y B G	Y R G R Y B	B G R G B Y	R Y B Y R B	R Y G Y R G	R G B G R B	Y G B G Y B

c. 30 different cubes can be made using six colours.

3D shape 7 page 69

1
a. 8 crates in a 6m container.
b. 16 in the 12m container and 24 in the two 9m containers: Total: 40.

2
a. Width: 5cm   Length: 10cm
b. App. width: 3.2cm   Length 6.4cm
c. App. width: 4.47cm   Length 8.94cm

3
a. With 40cm x 40cm, cut approximately 6.7cm to get maximum volume.
b. With 30cm x 20cm, cut approximately 3.9cm.

2D shape 1 page 70

1 11

2 1, 3, 6, 10, 15, 21, 28 . .

3 9 from 20

5 from 12

or

4 The true statement is that all squares are rectangles.

2D shape 2 page 71

1 There are five different tetrominoes and twelve different pentominoes.
(The pentominoes which are partly shaded are those which would fold to form an open box. The base is the shaded square.)

Tetrominoes   **Pentominoes**

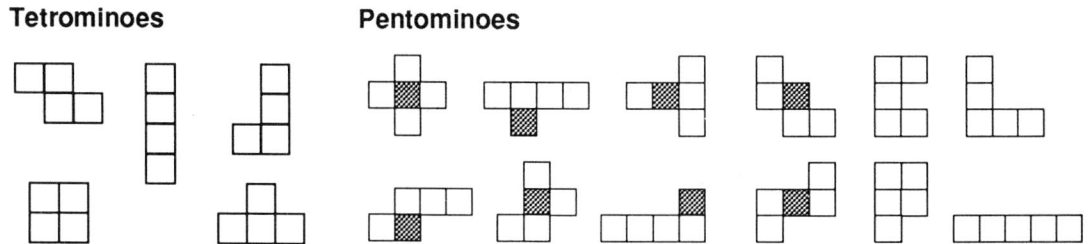

2 All the tetrominoes except the Z shape will tessellate to form a 4 x 4 square.

3 All the pentominoes will tessellate.

2D shape 3 page 72

1 Possible solutions are:

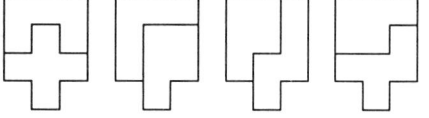

Only 2 will fit this shape.

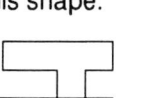

2

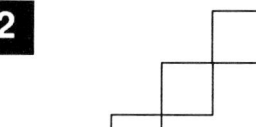

3 Here are the 35 different hexominoes. Those which can be folded to form a cube are shown shaded.

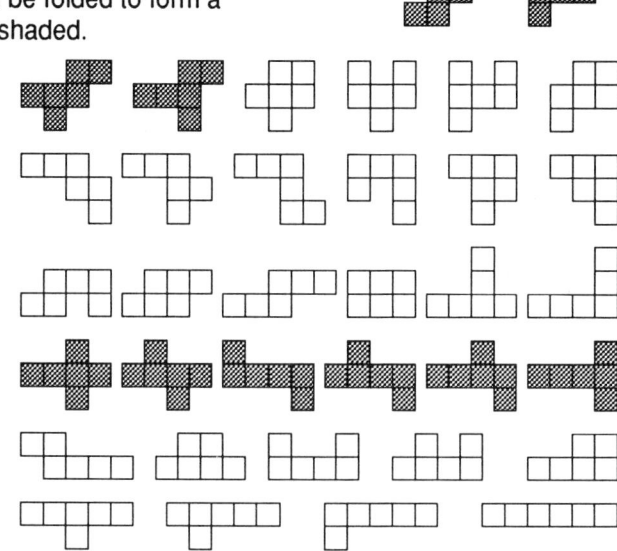

2D shape 4 page 73

1 8 squares with an area of 5 sq. units.

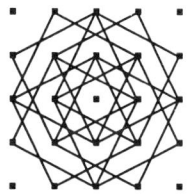

8 different sizes of squares.

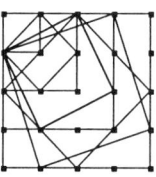

3 For example:

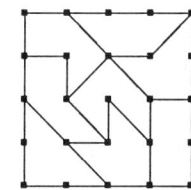

Total number of squares of all sizes is 50.

Area	1	2	4	5	8	9	10	16
No.	16	9	9	8	1	4	2	1

2 Area of largest triangle is 8 sq. units. There are 16 with rotations of these.

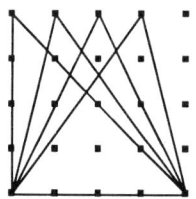

6 triangles with area $\frac{1}{2}$ sq. unit.

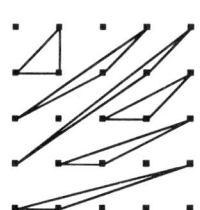

2D shape 5 page 74

1

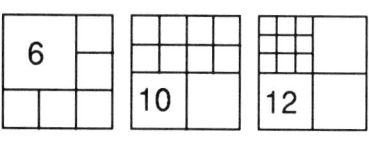

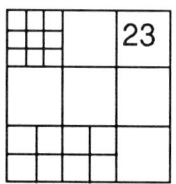

2

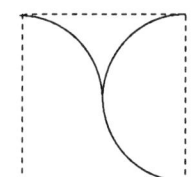

Only 2, 3 and 5 cannot be made.

2D shape 6 page 75

1 10 x π.

3 16 times each minute.

2
a. Each has the same amount of black.
If 2a is the length of the side of a square, white area in each pattern is πa^2.
b. Curved line is $4\pi a$ in E; $2\pi a$ in A, B and D; πa in C.

2D shape 7 page 76

1 12cm by 10cm by 8cm.

2 Rectangular areas of either 80cm x 55cm or 110cm x 40cm would do, with tape to seal edges.

2D shape 8 page 77

1 In addition to squares and rectangles, which are also parallelograms and trapezia, four parallelograms and five trapezia can be made.

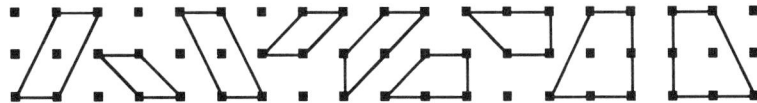

Many shapes with five, six or seven sides are possible on the 3 x 3 board.
Here are just some of those which will tessellate.

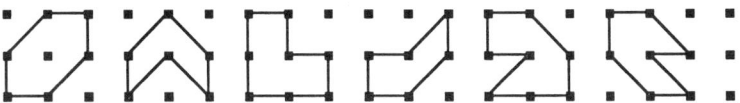

2 Large equilateral, small equilateral, large isosceles, small isosceles triangles.
Large square, small square, rectangle, various large and small rhombus, various parallelograms, kite, arrowhead, two isosceles trapezia.
With straws of three lengths, the scalene triangle and three different isosceles trapezia can be made, plus many more variations on the above shapes.

Symmetry 1 page 78

1
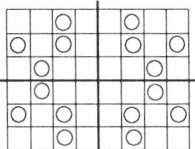

2
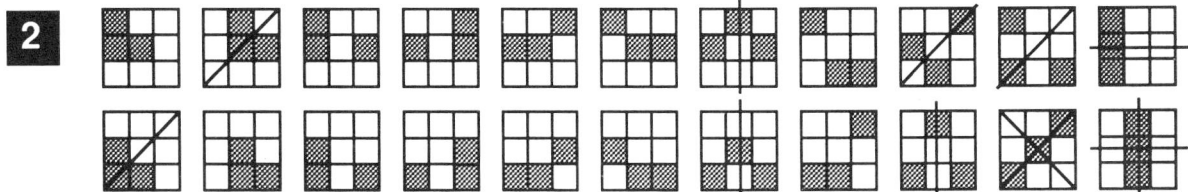

Symmetry 2 page 79

1
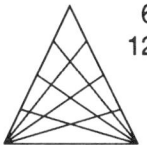
64 triangles
125 triangles

2

Symmetry 3 page 80

1 A, H, I, M, O, T, U, V, W, X and Y have a vertical line of symmetry and appear the same when reflected in a vertical line.
B, C, D, E, H, I, K, O and X have a horizontal line of symmetry and appear the same when reflected in a horizontal line.
If the letter O is drawn as a circle, and the arms of the letter X cross at right angles, these will appear the same when reflected in a sloping line.

2 H, I, N, O, S, X, and Z have rotational symmetry and appear the same after a half turn.

3 DEC 8.

Scale 1 page 81

1   a.   Sides of the black triangle are $\frac{7}{8}$ $\frac{5}{8}$ $\frac{3}{4}$ units.

       b.   Area of the black triangle is $\frac{1}{64}$ of area of the big triangle.

2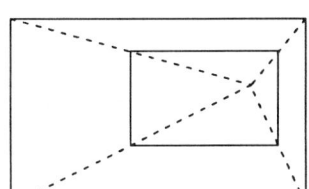

The point is at the centre of enlargement.

3   Increasing the angle turned, or decreasing the distance moved forward, will decrease the size of the circle: for example,

```
REPEAT 20 [FD 15 RT 18] or  REPEAT 36 [FD 5 RT 10]
```

Scale 2 page 82

1   One way to enlarge the boat is to trace it onto squared paper, then enlarge it on the same paper using a three to one scale. The smaller the squares on the paper, the more accurate the enlargement is likely to be.

2   The approximate height of the building is $\frac{48}{28}$ x 16 metres, about 27.4m.

Since the stick and its shadow were measured to the nearest centimetre, the length of the stick is between 47.5 and 48.5cm, and the shadow between 27.5 and 28.5cm. The shadow of the building is between 15.5 and 16.5m.

Greatest height: $\frac{48.5}{27.5}$ x 16.5 = 29.1m   Least height: $\frac{47.5}{28.5}$ x 15.5 = 25.8m

To find the height more accurately make sure the stick is vertical, measure at a time when shadows are longer, or measure to the nearest millimetre.

Angles 1 page 83

1   Dots which are a constant hopping distance from the cross will form a set of nested squares on the diagonal lines of the lattice.

2

Angles 2 page 84

1   There are 8 different triangles on a 3 x 3 pinboard, of which 4 have a right angle. There are 3 right angled isosceles, 1 right angled (not isosceles), 2 isosceles (non right angled) and 2 scalene triangles. It is not possible to make an equilateral triangle, but pupils should not be discouraged from trying. If different positions on the board are allowed, 76 triangles can be made.

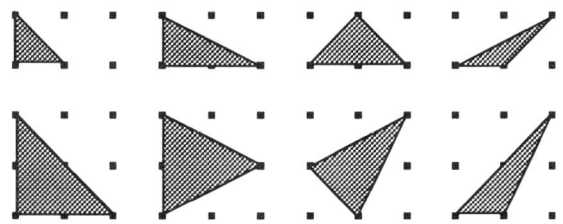

Angles 3 page 85

1   One hand between 3 and 4, and the other between 9 and 10: ie, 9.18 or 3.48.

2   a.   360°   b.   648°

3

No. of lines:	2	3	4	5	6	In general, n lines give n(n-1)
No. of angles:	2	6	12	20	30	angles.

Five lines equally spaced at 72° gives angles of 72°, 144°, 216°, 288°, 360°.

Angles 4 page 86

2

3 For example,

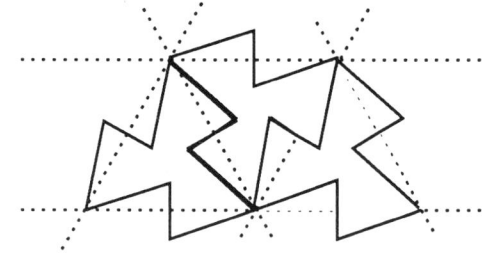

Bearings 1 page 87

1 a. i. 14 km   ii. 11.3 km   iii. 16.1 km   b. 39°   c. 116°   d. 128°

Networks 1 page 89

1   Five impossible.

2 Eg: ACEBDABCDEA.

3 The pencil must be taken off three times to draw the network.

4 Six more lines can be drawn: eg

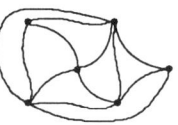

For three or more points (P), the most lines which can be drawn is 3(P-2).

Networks 2 page 90

1 Eg: ABJHIGFDCEA

2 Shortest route 19 miles: eg
BADGDEFIFCBEHKLIHGJK.

3 a. eg AGBHCDIEFA   360 metres
b. eg ABCDEFAGFIEIDHCHBGHIGA
840 metres.

Networks 3 page 91

1 16 routes: ABECD, ABEDC, ACDBE, ACEBD, ADCBE, ADEBC, BAECD,
BAEDC, BCDAE, BCEAD, BDCAE, BDEAC, ECABD, ECBAD, EDABC, EDBAC.

2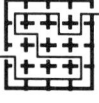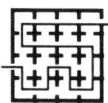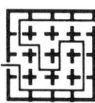

3 16 moves: A-E, C-G, G-A, H-D, D-C,
C-G, F-B, B-H, H-D, D-C, E-F, F-B,
B-H, A-E, E-F, G-A

Networks 4 page 92

1 Fastest trains are:
1106 from New Street to Reading
1031 from Coventry to Oxford
1107 from Oxford to Reading

2 Three distinct round trips:
A B C D E F A
A B D C E F A
A B D E C F A
making six altogether with reversals.
The last of these is the quickest:
total time 21 minutes.

Possible maps:

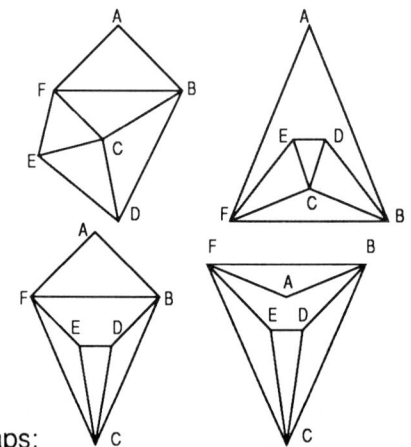

Data 2 page 94

1

	P	W	L	D	For	Ag
Reds	2	0	2	0	2	5
Greens	2	1	0	1	3	2
Blues	2	1	0	1	2	0

RvG: 2-3; GvB: 0-0; BvR: 2-0

2

Team	Played	W	D	L	W	D	Points	Place
Rovers	8	4	1	3	8	1	9	3rd
City	8	5	1	2	10	1	11	1st
United	8	4	2	2	8	2	10	2nd
Forest	8	2	2	4	4	2	6	4th
Spurs	8	2	0	6	4	0	4	5th

Data 3 page 95

1
a. beginning with p;
b. related to school;
c. having 10 letters;
d. ten letter words beginning with p.

2 At least four spaniel puppies.

Some other ways of grouping the words are: five letters; double s; beginning with d; containing a u; ending with . . . sion; meaning a person; and so on.

3

Data 4 page 96

1
a. Pea pods tested: 130
b. Most common no. of peas in pod: 4
c. Pods containing two peas: 15
d. Total no. of peas in pods with two: 30
e. Total number of peas: 495
f. Average = total peas/no. of pods = 3.8

Data 6 page 98

1

For example:		Multiple of 4:	F	Unlucky:	N
Odd:	C	Multiple of 5:	H	Triangular:	B, L
Even:	D	Greater than 8:	A	Square:	Q, M, P
Multiple of 3:	E, G	Half a dozen:	K	Cube:	J

Data 8 page 100

2 One way to estimate the number of letters is to fold or mark the square in quarters, count the letters in one quarter, and multiply by 4. Cutting a hole in a piece of paper to make a 'window', then placing the window in different positions, will reveal different or similar patterns. A useful size for a window would be 2cm x 2cm, but try windows of different sizes and shapes. Graphical representations of frequencies will help to show differences.

Data 9 page 101

1
a. Maria Pike 1774
    George Cook 1774
b. George Knapp 1875
c. 1857, 1859, 1870

2
a. 25 different surnames.
b. May.
c. John, George or Isaac.
d. Elizabeth.

3
a. 32 burials   b. 19 men (59%)   c. 13 women (41%)

4

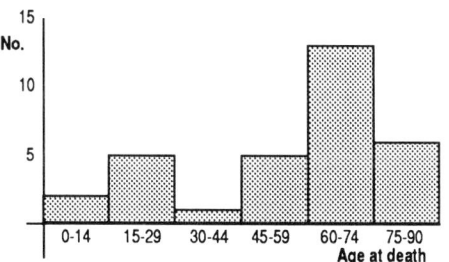

Male	2	3	1	3	7	3
Female		2		2	6	3
Age	0-14	15-29	30-44	45-59	60-74	75-90

Most common age at death: 60-74
Average (mean) age at death: 55

Probability of death before 60:
men 9/19, women 4/13.

Probability 1 page 103

2 The game is not fair. Evens can win 12 counters to every 9 won by odds.

Probability 2 page 104

1 Five socks to get a pair.

2 First guess: $\frac{1}{4}$ Last guess: 1

3 Only one cube need be examined. Pick one from the box marked RB, which must contain either RR or BB. If the cube taken out is red, then the box holds RR. The other boxes must contain BB and RB; BB can only be in box labelled RR, so RB is in box labelled BB.

Probability 3 page 105

1 12 different pies.

2 There are 7 possibilities for quins:
5 identical quins;
identical quads plus 1;
identical triplets plus identical twins;
identical triplets plus 2 non-identical;
two sets of identical twins plus 1;
identical twins plus 3 non-identical;
and all five non-identical.
For sextuplets there are 11 possibilities.

3
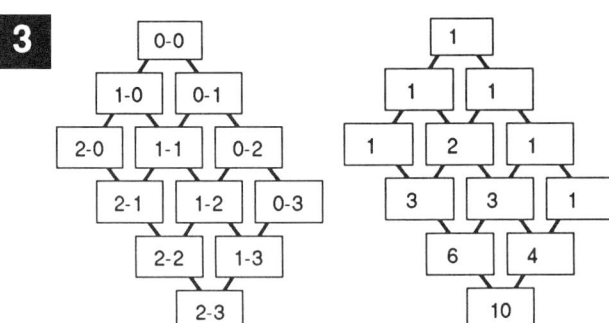

Final score could be reached in 10 different ways.

Probability 4 page 106

1 a. 6   b. 6   c. 12   d. 24

2 a. 4   b. 16   With three pens: a. 9   b. 81

3 There are 6 possible pairs of vertically adjacent squares and 6 possible pairs of horizontally adjacent squares. There are 36 ways of colouring two of the 9 squares on the grid, so the probability of two being adjacent is $\frac{12}{36}$ or $\frac{1}{3}$.

With a 4 x 4 grid, the probability is $\frac{24}{120}$ or $\frac{1}{5}$.

Probability 5 page 107

2

+	1	2	3	4	5	6	1st dice
1	2	3	4	5	6	⑦	
2	3	4	5	6	⑦	8	
3	4	5	6	⑦	8	9	
4	5	6	⑦	8	9	10	
5	6	⑦	8	9	10	11	
6	⑦	8	9	10	11	12	

2nd dice

3 Probability of 10 is $\frac{3}{36}$ or $\frac{1}{12}$.

The expectation would be five scores of ten in 60 rolls.

Probability 6 page 108

1 The game favours the second player who has 20 out of the 36 equally likely outcomes compared with 16 for the first player.

2 The best route in the obstacle race is 1b, 2a, 3b, 4b. At the first obstacle, the chance of 2 or 3 is $\frac{1}{12}$ and the chance of 7 is $\frac{1}{6}$, odds of 1:2 in favour of b.

The odds for each obstacle, taken in order a:b, are 1:2, 5:3, 3:4 and 5:9.